CHROMATOGRAPHIC
SYSTEMS

maintenance and troubleshooting

CHROMATOGRAPHIC
SYSTEMS
maintenance and troubleshooting

JOHN Q. WALKER
mcdonnell-douglas research laboratory
st. louis, missouri

MINOR T. JACKSON, JR.
monsanto central research department
st. louis, missouri

JAMES B. MAYNARD
shell research laboratory
wood river, illinois

ACADEMIC PRESS   NEW YORK AND LONDON   1972

ACADEMIC PRESS, INC.
111 Fifth Avenue, New York, New York 10003

United Kingdom Edition published by
ACADEMIC PRESS, INC. (LONDON) LTD.
24/28 Oval Road, London NW1 7DD

LIBRARY OF CONGRESS CATALOG CARD NUMBER: 72-75835

PRINTED IN THE UNITED STATES OF AMERICA

CONTENTS

PART I. LIQUID CHROMATOGRAPHY

PART II. GAS CHROMATOGRAPHY

CONTENTS

PREFACE

The purpose of this book is to provide a clear and concise guide for chromatographic maintenance—troubleshooting and repair procedures which can be utilized by both experienced and inexperienced chemists and technicians to reduce instrument down-time. No attempt is made to duplicate any of the excellent texts already published which may deal, in part, with troubleshooting chromatographic systems. Rather, we endeavor to go further by bridging the void between the chromatographer and the service engineer.

The text is divided into two parts. Liquid Chromatography, presented in Part I, consists of an introductory chapter on principles, techniques, and utility. This is followed by specific chapters devoted to the individual systems comprising the total liquid chromatographic makeup. The final liquid chromatography chapter, Comprehensive Troubleshooting, is provided for rapid reference. Part II, Gas Chromatography, begins with an introductory chapter on basic theory. This is followed by a systematic progression through possible malfunctions in various parts of the gas chromatograph, beginning with the sample introduction system.

The collaboration of three experienced chromatographers has definite advantages when considering diverse and complex subjects such as the maintenance and troubleshooting of chromatographic systems. The authors represent a total of 42 years of experience in several areas of chromatography and in different industrial backgrounds, i.e., aerospace, chemical manufacturing, and petroleum refining. For the past 10 years, each author has been involved in areas of chromatographic research with regard to the nonroutine applications associated with the needs of his employers. Included in these areas are those of control laboratory and analytical research supervision, plant start-up monitoring, and the usual day-to-day maintenance and troubleshooting problems. During our years of experience in various chromatographic functions, the authors have compiled careful notes regarding instrument malfunctions encountered in virtually every possible situation. Essentially, the text consists of these notes organized and rewritten in a readable, straightforward fashion.

The typical chromatographer who knows basic gas and liquid chromatographic theory and uses his equipment for many applications frequently considers the diagnosing and subsequent repair of instrument malfunctions something of an undesirable task and, in many cases, a black art. This need not be the situation because any adept chromatographer possesses the capability of demanding and receiving the maximum performance from his equipment. This capability is rewarding to him professionally and renders to him virtually full control and responsibility for the results attained. As a consequence, judgment regarding the validity of data is enhanced. In no context should this philosophy be construed that the chromatographer's primary goal be that of a service engineer. However, due to the advancements in chemical instrumentation technology, it behooves us to respect and appreciate some degree of inbreeding and/or overlap to fulfill our responsibilities as chromatographic scientists.

The operating cost of a $5000 chromatograph fully utilized for eight hours is approximately $45; however, the depreciation plus the overhead cost for the unit when not being fully utilized may well double the cost. This cost may be attributed to a reduction in the flow of analytical data, and, in many cases, less than full utilization of the analyst's time. A rapid diagnosis of the instrument or column malfunction with the appropriate remedy would substantially decrease instrument down-time and economize chromatographic maintenance.

Recent associations with various ASTM and API committees have shown that the gas chromatography expert must now be more proficient in the various types and capabilities of liquid chromatographic techniques in order to be a balanced separations scientist. So it is to this "balanced separations scientist" that we dedicate this book on the maintenance and the troubleshooting of chromatographic systems.

John Q. Walker
Minor T. Jackson, Jr.
James B. Maynard

ACKNOWLEDGMENTS

The completion of this book would have been impossible without the help and constructive criticism of many people, especially Dr. R. P. W. Scott, Dr. M. P. T. Bradley, Mr. W. K. Hinrichs, and Mr. J. L. McDonald.

Grateful acknowledgment is due our managers—Dr. D. P. Ames and Dr. C. J. Wolf of McDonnell-Douglas Corporation, Mr. D. R. Beasecker and Dr. W. E. Koerner of Monsanto Company, and Dr. J. W. Armstrong of Shell Oil Company— for their helpful suggestions and ready assistance.

We wish to thank Mrs. Eunice Crayne, Mrs. Loretta Goins, and Mrs. Joan Waid for their efforts in preparing the manuscript, and Mr. R. H. Pfeffer and Mrs. Katherine Sewell for preparing the artwork.

Finally, special thanks to our wives, Virginia Walker, Pat Jackson, and Susan Maynard, and our children for their help, encouragement, and understanding during the past year.

LIQUID CHROMATOGRAPHY

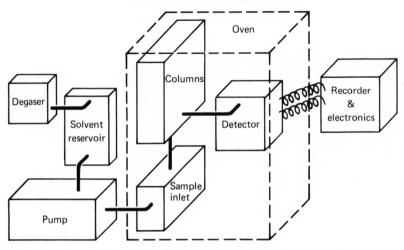

PART I. LIQUID CHROMATOGRAPHY

The general format of the liquid chromatography (LC) section of the text will be directed toward devoting specific chapters, in progression, to the individual systems comprising the LC make-up. The order in which these will be presented is as follows: Solvent Transport System, Sample Introduction System, Columns and Column Ovens, and Detection Systems. A final chapter on Comprehensive Troubleshooting is included also for rapid reference.

Chapter 1

"CHROMATOGRAPHIC SYSTEMS: MAINTENANCE AND TROUBLESHOOTING"

Introduction to Liquid Chromatography

This portion of the text, dedicated to high-speed liquid chromatography, is presented as a guide to enable the chemist to recognize, maintain and/or eliminate malfunctions which may occur within his particular instrument via preventive maintenance and trouble-shooting techniques. Figure 1-1 shows a block diagram of a typical liquid chromatographic system.

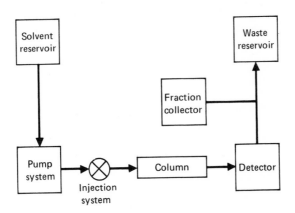

Figure 1-1   Block Diagram of a Typical Liquid Chromatograph

Liquid chromatography, the oldest form of chromatography, has been in existence for many years and few analytical techniques, if any, are as powerful or offer the potential magnitude of its application. Until recently the technique was time-consuming, requiring as long as twenty-four hours to perform a single analysis. Currently, the same analysis is being performed in minutes and therefore is deserving of its name, "high-speed liquid chromatography." This decrease in analysis time, with no subsequent loss in resolution, has been accomplished via efficient high-pressure pumping systems, increased column technology, high sensitivity detection systems and the overall reduction of chromatographic principles and knowledge to practice. Although much progress has been made, the present systems are far from optimum; and thorough knowledge of the technique and full awareness of its future potential cannot be completely visualized today. Virtually all parts of the current liquid chromatographic system are in need of improvement. It must nonetheless be realized that many problems associated with the present technology arise, at least in part, from insufficient knowledge and/or inadequate operation by the user.

There are excellent texts (1), (2), currently available devoted to theoretical and experimental considerations; however, some general background in liquid chromatography will be presented here relating to general principles and nomenclature. The four basic types of liquid chromatography may be categorized as follows:

1. Liquid-Solid or Adsorption Chromatography is an affinity separation associated with the separation of relatively nonpolar, hydrophobic materials. Normal phase adsorption consists of a non-polar solvent and a polar support. Silica gel and alumina are common choices using the solvents pentane and tetrahydrofuran, respectively. Silica gel is used to separate acidic or neutral compounds, whereas alumina is often used to separate basic compounds. Many workers note the undesirable effect of alumina in catalytically decomposing many organics; however, neutral aluminas are available which will completely eliminate or greatly minimize this problem. The solvent system is an equally important variable in adsorption chromatography. For a particular material, the more polar the solvent, the shorter the analysis time. Table 1-1, (3) consisting of an eluotropic solvent series, provides a convenient guide for solvent selection. Reversed phase adsorption has useful applications also. In this type of adsorption chromatography, a polar solvent and a non-polar stationary phase are employed.

Table 1-1
ELUOTROPIC SOLVENT SERIES (3)

Guide for Solvent Selection Listed in Order of Increasing Eluting Power

INCREASING ELUTING POWER

Solvent	Dielectric Constant	Temperature
n-Hexane	1.9	(20°C)
Cyclohexane	2.0	(20°C)
Carbon Tetrachloride	2.2	(20°C)
Benzene	2.3	(20°C)
Toluene	2.4	(25°C)
Trichloroethylene	3.4	(16°C)
Diethyl ether	4.3	(20°C)
Chloroform	4.8	(20°C)
Ethyl acetate	6.0	(25°C)
1-Butanol	17.0	(25°C)
1-Propanol	20.0	(25°C)
Acetone	21.0	(25°C)
Ethanol	24.0	(25°C)
Methanol	33.0	(25°C)
Water	80.0	(20°C)

Thin-layer chromatography is also a type of adsorption chromatography and has proven to be an invaluable technique for screening various solvents prior to instrumental adsorption chromatography. Speed and efficiency of analyses, in addition to easier quantitative analysis and preparative applications, render the instrumental technique more practical than conventional thin-layer chromatography.

2. Liquid-Liquid or Partition Chromatography involves two liquid phases (one mobile, the other stationary). Preferably, the two liquids are immiscible and exhibit different polarity. This technique is quite useful for isomer separations and can be used for virtually any separations involving components of the same or similar polarities. One inherent difficulty with this form of chromatography is that the mixture to be analyzed must exhibit solubility in the immiscible pair (mobile and stationary phases) chosen. Another operational problem has been minimized by the development of bonded liquid phases which render the stationary phase "more stationary" and eliminate the need to saturate the mobile phase with the stationary phase. Consequently, a wider choice of mobile phases becomes evident and also the latitude of sample - type applications. Prior knowledge of the nature of the samples to be analyzed is always necessary. Better success with liquid-liquid chromatography, as with liquid-solid chromatography, is realized by careful choice and pre-testing of the solvent systems.

3. Ion-exchange chromatography, comprised of macro- and micro-reticular resins and pellicular resins (see Figure 1-2) (4), provides separations on the basis of the ionic properties of the compounds to be separated. The process involves reversible exchanges of ions between the ion exchange phase and the components to be separated. The ion exchange process was one of the earliest chromatographic techniques used and has found its greatest application in the biological field.

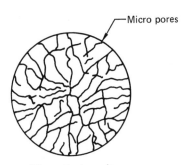

Microreticular resin

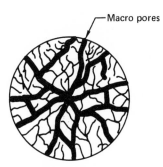

Macroreticular resin

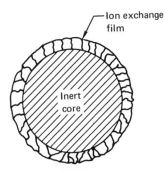

Pellicular resin

Figure 1-2   Structural Types of ION Exchange Resins
used in Liquid Chromatography (Courtesy of
C. D. Scott, Oak Ridge National
Laboratory)

Synthetic resins (cross-linked polystyrene) are
coated or bonded with an organic phase terminated
with anionic or cationic functional groups. Typical
functional groups are as follows:

Strong cation exchange = $-SO_3^-H+$

Weak cation exchange   = $-COO^-Na^+$

Strong anion exchange = $-CH_2N^+(CH_3)_3Cl^-$

Weak anion exchange   = $-N^+H(R)_2Cl^-$

The exchange capacity of the weak cation and anion resins is dependent upon pH over a reasonably narrow range; whereas, the strong cation and anion resins have exchange capacity over a much broader range and are, for all practical purposes, independent of pH within these limits.

Aqueous, low concentration buffered solutions are normally used; however, some applications demand mixtures of organic and aqueous solvents. The amount and nature of the organic solvent should be carefully selected, otherwise the coated organic phases may be stripped from the polymer bead and render the column useless for ion exchange applications.

In many cases, the gradient elution technique has drastically increased the efficiency of ion exchange chromatography. For example, many applications whereby the components to be separated are extremely sensitive to pH changes gradients may be established along the length of the column and achieve separations otherwise impossible.

In the analysis of unknown mixtures, frequently there are components which are irreversibly adsorbed on the resin. Prolonged usage under these conditions generally leads to poor column performance. This may be eliminated by using a short pre-column of the resin and periodically replacing it with a freshly prepared one.

Sample overload is often the cause for lack of resolution, especially when using small internal bore columns packed with pellicular resins. These columns can generally accommodate approximately 100 micrograms of component before overloading becomes evident.

Future developments in ion exchange chromatography will be directed toward increasing the speed of analysis. This will be accomplished, in part, through increased knowledge in the preparation of pellicular resins. Those pellicular resins which are permanently bonded to the inert sphere seem to provide columns of higher speed, efficiency, and longer column life in comparison with the "coated" sphere. The major difficulty encountered with

these resins is that of manufacturing processes.
Wide variations in performance have been observed
within different batches of the same resin. High-
speed preparative ion exchange chromatography should
be a most promising development in the near future.
Scale-up technology and the evaluation of resins
with new functionality will be critical areas for
research.

4. Exclusion (or) Gel Permeation Chromatography
is a separation of the components in accordance with
their molecular size. The heavier components,
exhibiting less permeation into the gel, are eluted
from the column first. This is probably the simplest
form of chromatography and is widely applied in the
field of polymer studies. Data is obtained in the
form of molecular weight distributions and for this
reason is very useful for surveying unknown sample
mixtures. A gel matrix provides the separation of
various molecular weight species depending upon the
size of the molecules to be separated and the pore
diameter of the gel particles. Some gels possess
active sites, and separations based strictly on
molecular size can be drastically altered by affinity
or steric effects.

A plot (Figure 1-3) of molecular size, in Angstrom

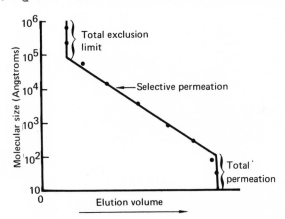

Figure 1-3   Ideal Gel Permeation Calibration Curve
              Showing Regions of Exclusion, Selective
              Permeation, and Total Permeation Obtained
              from the Plot of Log Molec
              Versus Elution Volume.

units, versus elution volume (mls.), shows the exclusion limit of the gel, whereby molecules of this
size exhibit no permeation into the gel pores and
are totally excluded. The exclusion limit is also
defined as the interstitial volume or the volume of
liquid in the column between the gel beads and is
available for removing from the column those molecules totally excluded by the gel. The total pene-
tration volume is defined as that portion of the
column available to smaller molecules. The region
of selective permeation is merely the difference
between the exclusion limit or interstitial volume
and total permeation and is the selective pore
volume of the column providing a plot of the elution
volume versus the logarithm of the molecular size.

The first objective in any chromatographic system
is that of obtaining adequate separations of the
components being sought. Separation is, of course,
dependent upon the resolving power (resolution, R_s)
provided by the column. Resolution may be defined
in three terms as follows:

$$R_s = 1/4 \quad (\frac{a-1}{a}) \; (\frac{k'}{1 + k'})(\sqrt{N})$$

Selectivity, Capacity , Random Dispersion

The relative retention,   , of a column for two components is dependent upon the ratio of the distribution coefficients, $\frac{k_1}{k_2}$, and describes the relative
selective retention of the components by the stationary phase. The capacity ratio, or factor, k', pre-
dominantly governs the length of time required to
obtain the desired separation. Wide var ations in
k' values for multicomponent systems defines the
general elution problem. k' is the ratio of the
amount of solute in the stationary phase versus the
amount of solute in the mobile phase. k' may be
altered by varying the eluant and the type or amount
of stationary phase. The random dispersion term,
N, relating band spreading and retention volume,
describes the fashion in which variables associated
with the number of plates, i.e., flow rate, column,
eluant, etc., alter the effects of other terms in
the resolution equation.

10

The Van Deemter equation, $H = A + \frac{B}{v} + Cv$, may be used to clarify further the effects of the random dispersion term in the resolution equation. The A term in the Van Deemter equation is the phenomenon of eddy diffusion or flow inhomogeneity. The B term is related strictly to band broadening (5) via the Einstein equation for diffusion. The C term is related to the forces of nonequilibria resulting from the resistance to mass transfer (6). Figure 1-4 is a plot of H versus v for a retained solute in both gas and liquid chromatography, where H is the height equivalent to a theoretical plate (HETP) and v is the mobile phase velocity. Optimum separations are those in which resolution is achieved in minimal time; therefore, the variation of H with v can be seen from:

$$t_R = N(1+k')\frac{H}{v}$$

where t_R (retention time) is directly proportional to $\frac{H}{v}$, considering that values for the N term (theoretical plate count) are contingent on R_s (resolution) and the (relative retention) value. The curves in Figure 1-4 for GC and LC are drastically different and these differences are attributed to the difference in diffusion coefficients between gases and liquids, the latter being 10^4-10^5 times smaller (2).

The theoretical considerations outlined above can only be optimized on a potentially productive liquid chromatographic system, i.e., the system must be properly maintained and the operator must be clearly aware of potential hazards and trouble-shooting tips to prevent costly downtime. Routine maintenance of the entire system such as periodic cleaning using solvents of different polarity is an excellent preventive measure for eliminating noise problems and unnecessary pressure drop. Inspection of all fittings as well as seals and gaskets within the pumping system may eliminate costly downtime and/or pump repair. A daily check with a suitable standard mixture provides a current assessment of column performance and changes in detector sensitivity. If one establishes a preventive maintenance program for his liquid chromatographic system, optimization of the three terms in the resolution equation is enhanced and additional time is realized for more productive pursuits.

11

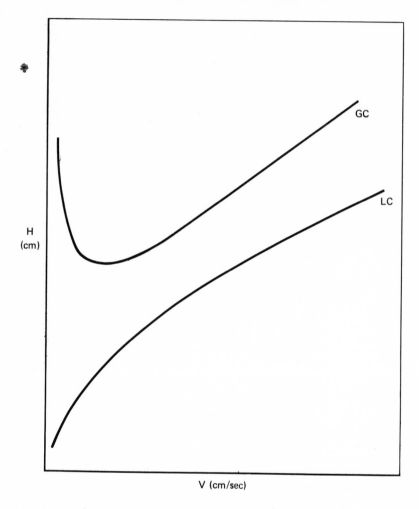

Figure 1-4 HETP Versus Mobile Velocity for GC and LC

References

1. L.R. Snyder, Principles of Adsorption Chromato-graphy, Marcel-Dekker, Inc., New York, 1968.

2. J.J. Kirkland, Ed., Modern Practice of Liquid Chromatography, John Wiley and Sons, Inc., 1971.

3. T.N. Tischer and A.D. Baitsholts, "Thin-layer Chromatography: Why and How", American Labora-tory, p. 72, May 1970.

4. C.D. Scott, "Ion Exchange Chromatography" Lecture Outline for Liquid Chromatography Course, Washington University, St. Louis, Missouri, 1970.

5. A. Einstein, Z. Elektrochem., 14, 1908.

6. J.C. Giddings, Dynamics of Chromatography, Marcel-Dekker, New York, 1965, p. 190-193.

Chapter 2

SOLVENT TRANSPORT SYSTEMS

The solvent transport system should efficiently accommodate the desired solvent and deliver it to the
various parts of the LC system. Many systems of this
type consist of: a degasser (for removing dissolved
air and other gases), a solvent reservoir (s), and
high-pressure pumps. Each part of this system will
be discussed in some detail.

Automatic solvent degassing systems offer time savings and more efficient operation. All aqueous
solvent systems and many organic systems should be
degassed. Manual degassing of solvents is effective
but quite time consuming. Manual degassing is accomplished by placing the solvent on a hot plate and
heating to a slow boil. The boiled solvent is then
cooled rapidly.

Generally, the use of stored degassed solvents, especially aqueous solutions, is not recommended because, on standing, sufficient air is re-dissolved
to render the degassing treatment ineffective. Dissolved gases in solvents have a tendency to produce
bubbles which may become localized in any part of
the system having sufficient dead-volume. Possible
dead-volume areas are the sample injection system,
column and detector. Some commercial degassing
systems are of the controlled vacuum type and are
quite efficient. Many of these provide auxillary
storage chambers designed for short-term storage,
i.e., for preparative work, when larger quantities
(recycle) of solvent are frequently needed. If a
conventional forepump or oil diffusion pump is used
as the controlled vacuum source, periodic checks regarding the oil level and condition of the oil in
the pump should be made. If the oil appears dirty,
a planned shut-down and oil change will eliminate
subsequent problems such as solvent contamination
and inefficient degassing.

Solvent change-over may cause many problems. These problems may be almost entirely eliminated by exercising a few precautions within the solvent transport system. Using miscible solvents (a carefully chosen series if necessary), thorough cleaning of the reservoirs for degassing and solvent storage, the pumping system, column (if possible), and detector before introducing the desired solvent is worthwhile. Many solvents off-the-shelf contain particulate matter which may be removed by filtration. Filtering of all solvents is recommended before degassing!

Efficient, high-pressure pumps are required to accomplish "high speed" liquid chromatography. These pumps must be of durable construction and all materials of construction must be as inert and corrosion resistant as possible. Metal parts are generally made of #304 and/or #316 stainless steel, the latter being preferable. The gaskets and seals are usually made of Teflon or Kel F. Many times solvent leakage occurs around the purge valve seats. This is often due to physical damage as a result of closing these valves too tightly. Slightly more than "finger-tip" tightening is required. If damaged, these valves may be easily replaced by the user. Additionally, a spare parts kit is available, upon request, from the manufacturer for practically all the pumping systems commercially available. It behooves one to have this on hand in case of pump failure. However, one should not be too eager to replace these components until other possible external causes have been eliminated.

Some possible external symptoms and causes are listed in Table 2.

The most widely used types of pumps for high-speed liquid chromatography may be broadly categorized as either "pulsed" or "pulseless." Many of these pumps are capable of high pressure operation, to 5000 psi, if needed. Except under the most extreme conditions 3000 psi capability is probably adequate. Pressures from 1000 to 3000 psi are generally required for small particle size, narrow bore column applications and provide rapid, efficient analyses.

Table 2

External Causes Erroneously Associated with Pump Malfunction

Sympton	Possible Cause and Remedy
1. Pump Fails to Build-up Pressure.	1. Solvent resevoir empty 2. Improperly closed valve (s) 3. Defective solvent line or connection 4. Air in pump, may require priming 5. Oil level too low for maximum efficiency 6. Leaking seals and gaskets. (Refer to specific pump manual for replacement)
2. Pump Pressure High Enough but No Flow to Injector, Column or Detector.	1. Pressure build-up due to excessive pressure drop; check for plugging in transfer line to injector, column, and detector. Flow is probably escaping through relief valve 2. Leak in system; check all fittings, septa etc.
3. Noisy, Erratic and/ or Pulsating Recorder Trace.	1. Pulsed Pumps - Underdamped, should incorporate additional pulse-dampener. 2. Air bubbles entering detector - Solvents and/or column needs additional degassing or purging. 3. Solid, particulate matter from column entering detector. Detach column and back-flush detector and transfer line from column

Sympton | Possible Cause and Remedy

(No. 3 continued)

        to detector.

4. Previous solvent not completely removed from system. Use step-wise miscible solvent purging if consecutive solvents are immiscible.

5. Detector leaking eluant. Remove covers and inspect for leaking seals or damaged cells.

6. Recorder problems:
   a. Dirty slide wire- clean with aerosol slide wire cleaner and/or methanol.
   b. Weak tubes in amplifier circuit.
   c. Mechanical binding of pen and/or slide wire.
   d. Inadequate connection of signal leads to recorder.
   e. Grounding problem.
   f. Gain adjustment too high.

4. During Operation, Pressure in Pump:
   (a) Increases

1. Increases in pressure during operation is indicative of plugging somewhere in the system. Suspect locations are: Pulse dampeners sample injectors (especially if septum injection is being used), column, detector.

   (b) Decreases

2. Decreases in pressure during operation is positive indication of a leak in the system. Check points are: septa, "O" ring seals in loop injectors, all fittings, cell (cell windows or spacers may be defective).

Pulsed pumps generally employ a single plunger multiple-
stroke operation (see Figure 2-1 and 2-2) with an

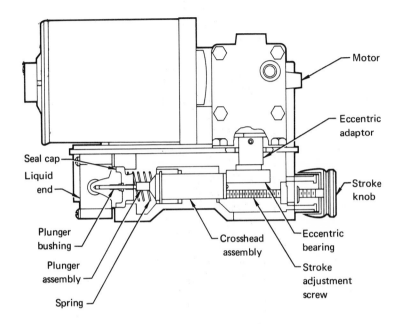

Figure 2-1 Typical Pulsed Pump used in Liquid
Chromatography (Courtesy of the Milton
Roy Company)

electronic pulsing circuit; however, pumps are available in the triple piston design whereby a continuous
flow of solvent is provided. Single plunger multiple-
stroke pumps require incorporation of pulse dampeners
or accumulators between the pump and sample injector
for eliminating a pulsating recorder trace caused by
the plunger strokes. The triple piston design more
closely approaches the performance of a pulseless
pump than other piston-driven pumps of this type;
however, the pressures attainable with these pumps
is somewhat less than the current single piston design.

Pulseless pumps are of the positive mass displacement syringe design. This pump consists of a reservoir and a single-stroke piston. The eluant in

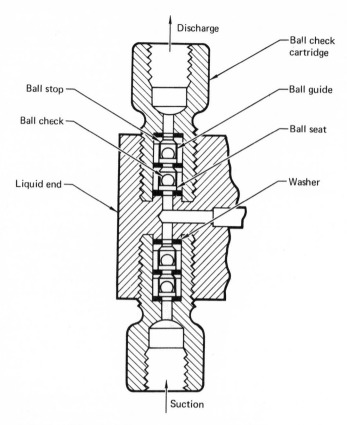

Figure 2-2 Liquid End of Pulsed Pump (Courtesy of
the Milton Roy Company)

the reservoir is gradually but continuously displaced
by a single piston-stroke operation, see Figure 2-3.
Flow rates are controlled by an electronic selector
switch. As with pulsed pumps, flow rate reproducibility is very good (values range from ±0.1% to
±1.0%) for the syringe-type pumps. The pulseless
(single piston displacement) pumps have O -ring
seals surrounding the periphery of the piston and
care should be taken against etching or marring these.
Small leaks can develop whereby quantities of the
eluant to be displaced may actually migrate behind
the force of the piston and cause inadequate delivery.

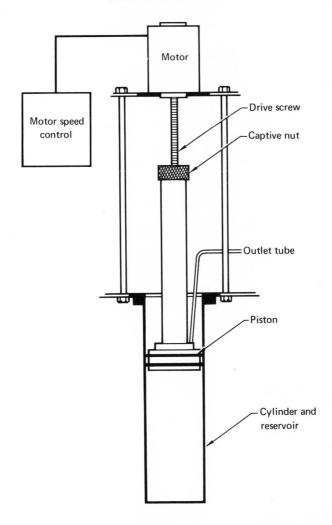

Figure 2-3   Pulseless Pump used in Liquid Chromatog-
            raphy (Courtest of the Nester Faust
            Manufacturing Corporation)

The type of pump chosen for a liquid chromatographic
system is virtually independent of the nature of the
problem to be solved provided adequate pressure can
be developed (~3000 psi) and measures for minimizing the effects of corrosive materials are taken.
Instrument manufacturers and suppliers are designing

their LC units more and more around the pulseless
variety of pumping systems. The pulseless pumps aid
in achieving higher sensitivities from the detectors
because the signal to noise ratio is higher. More
recently, extremely rapid solvent refill systems
have been made available for the pulseless, single
piston-stroke pumps. In the past, this was a
sought-after feature because the need to dismantle
the reservoir from the pump for refilling resulted
in loss of time and lessened efficiency of operation.
One recommendation is that two of these pulseless
pumps be readily available for maximum capability.
Automation of solvent refill systems to permit
continuous, uninterrupted analysis time for around-
the-clock operation then becomes feasible. This
continuous operation is more attractive and practical
with pulsed systems and requires only one pump;
however, automatic reservoir refill is necessary.
Gradient elution techniques, probably the most useful
aspects of liquid chromatography, is possible with
either pulsed or pulseless systems. The practical
state-of-the-art in gradient elution seldom requires
more than two eluants in varying proportions to
accomplish the desired separation.

Cleaning or washing of the pumping system is always
necessary when the operator desires to change eluants,
especially if the eluants have different polarities.
It is not necessary to physically disassemble the
pump to do this. A miscible solvent series used
in a step-wise manner insures thorough cleaning of
the pump. Adequate purging with the ultimate solvent
of choice is mandatory. All valves, fittings and
transfer lines associated with the pump should likewise be cleaned, preferably simultaneously with pump
cleaning.

Solvents known to be detrimental to proper pump
function are strong acids, especially halogen acids,
with pH 2.0 and extremely strong bases. Even short-
term use of the above solvents may rapidly cause
corrosion and undesirable deposits in valves, valve
seats and virtually any metal surface in the pumping
system.

Restrictors or pulse dampeners are sometimes used
in pulsed pumps for lowering the noise level observed
on the recorder trace due to the pump stroke. These
devices are generally positioned between the pump

exit and the sample injector. Particulate matter in the solvent emerging from the pump may cause plugging of these restrictors with subsequent pressure build-up. These may be removed from the system quite easily and be replaced with little time required for equilibration. An obvious symptom of this is excessive pressure build-up with no increase in flow rate, as measured at the detector exit. The increase in flow is occuring and being delivered by the pump except the direction of the flow is through the pump relief valve. During normal, optimum operation there should be no solvent escaping through this relief valve unless it has experienced some damage, or pressure build-up becomes excessive. It is recommended that a spare relief valve or a maintenance kit be kept on hand.

Concluding, virtually any part of the solvent transport system can be rapidly maintained by the operator provided an adequate supply of fittings, valves and seats are conveniently available. The time and capital investment for replacement spare parts is extremely economical compared to an instrument supplier's service call.

Chapter 3

LC SAMPLE INTRODUCTION SYSTEMS

The sample introduction system in liquid chromatography enables the operator to introduce his sample
efficiently into the system without disrupting the
established flow equilibrium in the column and detector. This criterion may seem easy to achieve;
but, only a few of the current designs adequately
fulfill this requirement.

The types of sample introduction systems are septum
injectors, stopped-flow injectors, and loop inject-
tors. Each of these will be described together with
the inherent advantages and disadvantages of each.
Ideally, the system of choice should introduce only
minimal dead-volume to the system, lest column efficiency be sacrificed

Septum injectors, probably the most widely used
technique, permit the operator to introduce his sample, virtually on-column, into the system. As in
gas chromatography, this seems, theoretically, to be
an extremely efficient technique for maintaining and
optimizing the theoretical plate-count of the column.
Septum injectors, of proper design (Figure 3-1),
(1), do provide capabilities for minimizing dead-
volume in the system. Normally, 5-50 μl syringes
are used for introducing the sample. This technique
is quite useful in the 10-500 psi range of pressure
delivered by the pumping system; however, at pressures above 500 psi problems become evident. Possibly, the major limitation is the choice of septum
materials, 1) for withstanding the high pressures in
the system and, 2) compatible with a wide range of
solvents. The more commonly used septa and recommended solvent applications are seen in Table 3-1.
This table is presented because the improper combination of septum and solvent can produce undesirable
results. Major problems encountered are those of
septum/solvent interaction by "leaching" materials
(generally plasticizer) from the septum and hardening

of the septum limiting its useful application to one
or two injections before it begins leaking. The
"leaching" phenomenon will become quite evident and
is a perfect analogy to gas chromatography septum
"bleed" as observed on the recorder trace. A gradual
increase or "drifting baseline" will be the result.
Because the detector has a low dead-volume this added
signal will result in a decrease in the sensitivity
of the components being sought in the analysis. The
linear range of the detector for the components of
interest will also be affected unless accommodations
are made for the lower sensitivity. Subsequently,
septum/solvent interaction results in hardening of
the septum. In many cases, the septum actually becomes brittle and even without the enhancement of
syringe injection may begin leaking, especially at
high pressure. Disruption of the flow characteristics
within the system are inherent with the changing of
the septum and, generally, re-equilibration of the
system takes longer than for gas chromatography. In
changing the septum, precautions should be taken to
insure that only minimal (if any) air is introduced
into the system as this will merely prolong equilibration time. This is best done by reducing the flow
of solvent delivered by the pump. Do not stop the
flow of solvent completely! Remove septum and adaptor,
change septum and slowly re-mount the adaptor, allowing excess solvent to flow into the adaptor and displace any air which may become entrapped. Then
tighten the adaptor and resume normal operation.

The technique of syringe injection also merits a
few comments. A fixed-needle type of syringe is
recommended. The needle should be tapered and free
of barbs, otherwise the useful life of the septum
is decreased. After filling the syringe and before
injecting the sample, any air entrapped in the
syringe should be removed. This can be done by inserting the needle of the "loaded" syringe into a
piece of silicone gum rubber and slowly depressing
the plunger to expel 5-10 μl for preferential removal of air.

Stopped-flow or interrupted flow injection is becoming more widely used as a sample introduction
technique. It is best used, however, with pulseless
single-stroke syringe-type pumps. Designs are available which prevent virtually any disruption in the

system and with minimal introduction of air, providing the preceding precautions on syringe filling
have been employed. Syringe injection is used
exclusively. Pulseless pumps are preferably used
because for stopped-flow applications these seem
to re-equilibrate more rapidly than pulsed pumps.
For non-routine applications the stopped-flow technique is best utilized by incorporating an internal
standard with the mixture to be analyzed. This is
especially true if quantitative data is being sought.
The use of a stopped-flow injector reduces dead-
volume in the system.

Loop injectors, which are used in almost all high
molecular weight GPC (Gel-permeation Chromatography)
applications, are becoming widely used in high resolution liquid chromatography also. Two features
of these injectors had to be improved for high resolution work. The loop injectors used for GPC
work had too much dead-volume for high resolution
LC use. Secondly, these type injectors have to
operate over a wider range of pressures. Currently,
there are several suppliers of these loop injectors
which are compatible with high resolution LC demands.
Perhaps the weakest links in any loop injector
available are the "O" ring seals incorporated in the
design. However, the more recent designs, (Figure
3-1) (1) are quite durable and are capable of ac-
commodating pressures to   2000 psi without leaking.
The principle involved in the loop injector is that
of syringe-filling a small column loop with sample
and, upon sampling, all or a portion of the sample
in the loop is displaced by diverting a portion of
the mobile phase into the loop. Sample loops of
5-10 ul are ideal for high resolution LC in contrast
to the normal 2cc loops used for larger column GPC
applications. Ideally, the sample to be analyzed
should be dissolved in portions of the solvent used
as the mobile phase to eliminate unnecessary solvent
peaks. For high resolution LC, the loop injector
seems superior to the other types of injectors.
Design of current loop injectors has been toward low
dead-volume, high-pressure capability. These features are required when narrow-bore columns containing small particle-size packings are used. Another high
pressure sample valve is shown in Figure 3.2.
The sample injection system must be kept clean. Again,
this is best accomplished via miscible solvents and

27

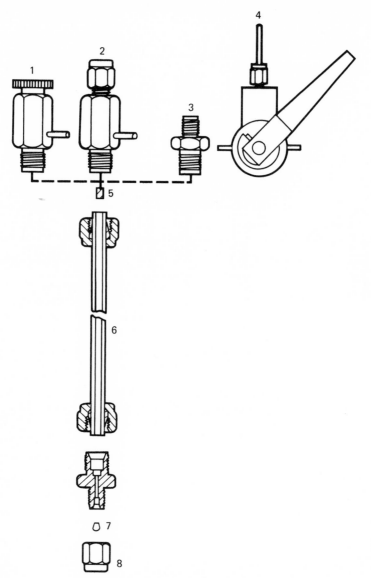

Figure 3-1   Sample Injector Designs for Liquid
Chromatography (1) On-column septum injector (2) Stop-flow injector (3) Reducing union with bonded frit (4) Loop
injector (5) Teflon filter retainer
(6) Precision-bore, heavy wall column,
316SS (7) Silver-plated ferrule (8)
316SS Nut (Courtesy of Nester Faust
Manufacturing Corporation)

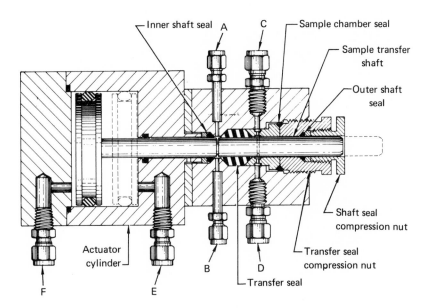

Figure 3-2 High Pressure Sample Valve for Liquid
Chromatography usable at pressures to
5000 psig. Connections are: (A) Mobile
phase, (B) Chromatographic column,
(C) Sample, (D) Drain, (E) and (F)
Solenoid actuated 2-way pneumatic valve.
Air source is 50-100 psig. (Courtesy of
Hamilton Company)

may be carried out simultaneously with pump cleaning. The use of corrosive solvents should be eliminated as most injectors will be damaged by such treatment.

Occasionally, septum injectors will become clogged after repeated injections. Generally, this plugging is due to small particles of the septum being deposited in the injector. Reversal of flow through the injector will remove the plug.

Concluding, the loop injector seems preferable for high resolution LC applications because current designs achieve low dead-volume, and flow equilibrium is disrupted less with this type injector. Additionally, virtually no air is introduced into the system. Improvements in all areas concerned with sample injection are forthcoming and will contribute to optimum development of high resolution liquid chromatography systems.

References

1. Septum Injector, Nester Faust Mfg. Corporation.

2. High pressure sample valve, Hamilton Corporation

Table 3-1

Recommended Choice of Septa for Use with Various
Solvents

Solvent	Recommended Septum
Methylethyl Ketone	EPR
Tetrahydrofuran	EPR
Dimethylformamide	BUNA-N
Alcohols	BUNA-N
Toluene	Viton-A
Benzene	Viton-A
Trichlorobenzene	Viton-A
Cresols	Viton-A
n-Hexane	BUNA-N or Viton-A
Water	EPR, BUNA-N, or Viton-A
Most Solvents	White Silicone Gum Rubber

Chapter 4

LC COLUMNS (SELECTION & PREPARATION), COLUMN OVENS AND COLUMN HEATERS

Selecting the proper mode or method of liquid chromatographic separation is not always straightforward.
Table 4-1 provides a convenient guide for establishing the type separation required for most systems
encountered in liquid chromatography. A more fundamental description of the various type (liquid-liquid,
liquid-solid, ion exchange, and gel permeation) is
found in Chapter 1, "Introduction to Liquid Chromatography."

Stainless steel or glass columns are commonly used.
For pressures exceeding 500 psi metal columns must
be used. Currently available glass columns and associated fittings have a quoted pressure limitation
of 500 to 1000 psi. Karger and Barth (1) have concluded that there were only small differences in
column efficiencies for SS, Al, or Cu glass columns;
whereas, teflon-coated aluminum tubes were found to
yield efficiencies that were poor and non-reproducible.
In their study, they also concluded that wall effects
(active sites and roughness) were insignificant for
the systems evaluated.

The diameter and length of the column chosen is
governed by the mode and efficiency of separation
required. For gel permeation chromatography, standard-wall 3/8 inch o.d. or 1/4 inch o.d. stainless
steel or glass columns are used. (pressures seldom
exceed 500 psi for GPC applications.) For high
resolution liquid-liquid, liquid-solid, or ion-
exchange applications where the pressure drop along
the column is very high, 1/8 inch o.d. (narrow-bore)
or 1/16 inch o.d. (thin-wall) stainless steel tubing
is used. Column lengths for GPC applications are
necessarily longer than those needed for other
modes of liquid chromatography. Effective column
lengths for GPC applications equivalent to sixty
feet are not uncommon especially if the solvent

Table 4-1

CHOOSING THE CORRECT MODE OF LIQUID CHROMATOGRAPHY

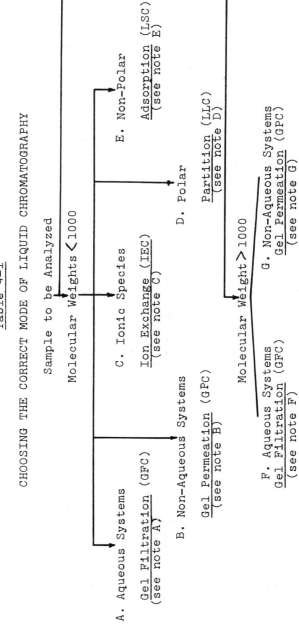

Sample to be Analyzed

Molecular Weights < 1000

Molecular Weight > 1000

A. Aqueous Systems

Gel Filtration (GFC)
(see note A)

B. Non-Aqueous Systems

Gel Permeation (GPC)
(see note B)

C. Ionic Species

Ion Exchange (IEC)
(see note C)

D. Polar

Partition (LLC)
(see note D)

E. Non-Polar

Adsorption (LSC)
(see note E)

F. Aqueous Systems

Gel Filtration (GFC)
(see note F)

G. Non-Aqueous Systems

Gel Permeation (GPC)
(see note G)

Note A

Aqueous Systems

Gel Filtration (GFC)
1. Choose smallest mesh possible to attain high efficiency.
2. Choose correct porosity according to molecular weight range of interest.
3. Avoid excessively high pressure.

Note B

Non-Aqueous Systems

Gel Permeation (GPC)
1. Column materials recommended:
   a) cross-linked polystyrene
   b) polyvinylacetate
   c) porous glass
2. Solvents
   a) If UV detection is used, choose a UV transparent solvent.
   b) For maximum sensitivity when using RI detectors, choose solvents which reflect the maximum RI compared to the species being separated.

Note C

Ionic Species

Ion Exchange (IEC)
1. Decide which type of resin to use, i.e., anion or cation.
2. Always use resin of smallest particle size to obtain maximum efficiency. For analytical scale work, use pellicular resins.
3. Other than column choice the most important variables are pH and ionic strength of the eluant. For anions pH 7, for cations pH 7 Ionic strength is increased to elute more strongly bound species.
4. Gradient elution is quite useful for optimizing the system

Note D

Polar

Partition (LLC)

1. Must choose a non-solvent for the liquid phase or the solvent must be saturated with liquid phase,
2. If a bonded liquid phase is used, solvent programming may be employed.
3. UV detectors may be used to eliminate drift if a low optical density solvent is used. The species sought will show negative adsorption.

Note E

Non-Polar

Adsorption (LSC)
1. Use TLC solvent screening.
2. If more than one solvent is needed to obtain the separation and RI detection is used, choose solvents whose RI is approximately the same.
3. If the species are UV adsorbers and a UV detector is being used, choose solvents which are transparent in the UV region. Be aware of UV adsorbing inhibitors which are added to non-UV adsorbing solvents.

Note F

Aqueous Systems

Gel Filtration (GFC)
1. Use low porosity gels. (See Aqueous Systems, Molecular weights >1000)

Note G

Non-Aqueous Systems

Gel Permeation (GPC)

1. Use low porosity gels. (See Non-Aqueous Systems, Molecular Weights >1000)

transport system (see Chapter 2) has recycle capability. LLC, LSC and ion-exchange chromatographic column lengths are generally two to ten feet in length.

The shape or geometry of the packed column has received much attention in correlation with column efficiency. Various configurations have been used (straight, round, square, etc.) and again seem to be quite dependent upon the particular mode of LC being employed and also the efficiency being sought. Straight columns are preferable because the homogeneity obtained during packing is preserved in use. Additionally, better reproducibility can be obtained among columns which are packed straight and used as such.

Many column compartments will accommodate only certain lengths of columns. Several straight columns may be used in series in such cases by "coupling" the column ends with short (one to three inches) lengths of small diameter, low dead-volume tubing. One-sixteenth inch (o.d.) and three to four one-hundredths inch (i.d.) stainless steel tubing is used. The fittings associated with these couplings should always be of low dead-volume design. In most cases, portions of the packing material are placed into the dead-space of the union eliminating the void area. The tube coupling is semi-circular in shape.

Pre-packing with subsequent coiling and/or bending of longer columns has provided satisfactory results. Configurations of this type are used with one-eighth inch (o.d.) narrow-bore (1-2 mm i.d.) or one-sixteenth inch (o.d.) thin-walled capillaries. Some workers (2,3) claim that round and sharp bends in LC columns have advantages by decreasing the resistance to mass transfer of sample components passing through the column. Presumably, small voids and packing inhomogeneities are formed within the column which are virtually impossible to reproduce. If these columns are not excessively long, the overall efficiencies may be poor, yet reproducible, within certain limits. There are no advantages realized in reproducing poor efficiencies. For example, in using a narrow-bore column ten feet in length having two or three sharp bends and correspondingly two or three voids, it follows that column efficiency may be drastically impaired. In spite of

37

the advantages obtained via decreasing the resistance
to mass transfer, voids contribute to eddy currents
in the column with subsequent recombinations of separated components and band broadening which, for
high resolution liquid chromatography, is intoler-
able. One could undoubtedly achieve the same or better efficiency using a straight column three feet in
length.

Methods of packing columns for liquid chromatography
vary widely depending upon the nature of the material to be packed. Most columns for liquid-liquid
or liquid-solid chromatography are merely dry-packed
using tamping or a mechanical vibrator as in gas
chromatography. Pellicular ion-exchange resins may
also be dry-packed. Conventional ion exchange resins must be wet-packed (4) in the form of a slurry,
allowing adequate time for swelling and particle
distribution between each addition of the packing.
It is recommended that once these columns are packed
in a wet slurry they should be kept this way when
not in use by capping the ends of the column.

GPC columns are pressure-packed in the form of a
slurry also (5). Again, as with conventional ion-
exchange resins, once packed, should not be per-
mitted to go dry. If the columns lose liquid and
become dry, contraction of the gel occurs creating
voids in the rather large diameter column. Subsequently, the column is virtually ruined because even
replenishing the liquid lost or the addition of more
packing material does not restore the column to its
initial efficiency.

GPC columns are generally purchased pre-packed from
the manufacturer. These columns are quite expensive
and rightfully so because few workers in the field
can reproduce GPC columns packed in their own laboratories. For this reason alone, it is recommended
that GPC columns be purchased pre-packed. This has
its advantages in spite of the costs involved.
Generally, the supplier will provide information regarding the efficiency or plate-count for any given
set of columns purchased. If, when using these
columns in your own laboratory, one does not reproduce the quoted efficiency, the columns may be
returned to the supplier either for credit or replacement columns. The theoretical plate count

varies in accordance with the packing material,
column diameter, and performance tolerance of the
column. When ordering GPC columns, provide separation only for the molecules you wish to separate.
Do not specify columns with larger exclusion limits
than the limit required by the largest molecules you
wish to separate. Most GPC columns ordered have a
specified capability of molecular weight applications.
Generally, this capability is expressed in Angstrom
units, a rather arbitrary designation. Even so, this
designation has useful application within a factor
of two, regarding molecular weight. With this in
mind, then, the GPC Angstrom size of a molecule is
the molecular weight divided by 20. Generally, molecules which have a high density per unit molecular
length, such as polystyrene, have a higher multiplier, and molecules which have a fairly low density
per unit chain length, such as polyethylene or poly-
propylene, have a multiplier as low as 15. The GPC
Angstrom size is defined as 1/41 of the molecular
weight of polystyrene (6). Some things to keep in
mind when evaluating GPC columns for efficiency and
reproducibility are:

a) a change in the volume of the system between
the injector and the column.

b) a change in the volume of the system between
the outlet of the column and the detector.

c) a change caused by the density of packing in
the column (different solvents, etc.).

d) a change caused by the density of the pores
and pore-size distribution within the polymer head, i.e., different solvents, pressure, etc.

Each of these changes, individually, are small; however, a combination of any of these can result in
considerable deviation from any standard curve.
It behooves one, then, to evaluate his particular LC
system and be aware of possible anomalies which may
occur.

In contrast to the earlier thinking of some workers,
column heating and column ovens drastically influence the efficiencies of some applications in

liquid chromatography. For GPC applications the effects are minimal; however, for high-speed liquid chromatography the results can be remarkable. This is especially true for high-speed ion-exchange liquid chromatography whereby the ion-exchange process is remarkedly enhanced by additional heat applied to the column. The detection system used should be compatible with the additional heat applied. There are column ovens and column jackets. As in gas chromatography, cold spots in the column or cold spots between the column exit and the detector may produce undesirable and non-reproducible results. The ideal approach, applicable to all detection systems, is that of a thermostated column oven, whereby the column and detector may be maintained at approximately the same temperature. The output of a differential refractometer is obviously temperature dependent, whereas an ultra-violet detector yields only minimal response to small changes in temperature. Column jackets, whereby column temperatures are controlled very accurately via circulation of water through a jacket, are very efficient and are used quite extensively with UV detectors; however, when used with differential refractometers are quite undesirable because of baseline drift. This happens because the small unheated dead-volume between the heated column and the refractometer fails to maintain the desired temperature. Application of heating tapes to this dead-volume area does not produce completely satisfactory results. The incorporation of dual-columns alleviates the situation somewhat regarding the refractometer detector output; however, one must be sure that flow effects due to temperature changes between the two columns are quite similar. Non-uniform heating within these regions contributes to baseline drift and inadequate detector sensitivity.

Concluding, overall, liquid chromatographic separations will improve almost proportionally to the user's knowledge and experience with his particular unit with regard to choice of separating modes, proper column choices and the use of controlled and equilibrated column heating. In the liquid-liquid chromatography mode, column heating is especially significant. All applications in this area rely upon the compatible solubility of components to be separated in the liquid or stationary phase. This solubility

is inherently temperature dependent. Virtually all
of the liquid-coated inert spheres have maximum
temperature limitations and these limitations should
be adhered to in-practice. The use of a pre-column
or an eluant containing a portion of the coated
liquid phase is essential for replenishing the loss
of liquid phase when a sufficiently high temperature
is used on the column.

References

1. B.L. Karger and H. Barth, Analytical Letters,
    Vol. 4, No. 9, September 1971, p. 602.

2. R.P.W. Scott, D.W. Blackburn, and T. Wilking,
    J. Gas Chromatography, 5, 183, 1967.

3. J.J. Kirkland, J. of Chromatographic Science,
    7, 361, 1969.

4. C.D. Scott and N.E. Lee, J. of Chromatography,
    42, 263, 1969.

5. K.J. Bombaugh in Modern Practice of Liquid
    Chromatography, J.J. Kirkland, Ed., John Wiley
    and Sons, Inc., New York, 1971.

6. Chromatography, Waters Associates, September 1970.

Chapter 5

LC DETECTION SYSTEMS

The general purpose of detection systems in liquid chromatography is evident. The choice of a suitable detector, many times, is not quite as evident. The liquid chromatographer has at his disposal numerous means of detecting what he has so laboriously separated. Therefore, the choice of detection is important to the overall scheme of the LC make-up and is contingent upon such criteria as the noise, sensitivity and linearity requirements of the problem.

The two basic categories of detectors are solute property detectors and bulk property detectors. The solute property detector is sensitive to some physical property of the component being analyzed and is relatively insensitive to the eluant or mobile phase. Examples of this type are UV adsorption, fluorimeters, radioactive detectors, solute transport detectors, and polarographic detectors. Bulk property detectors monitor changes in the physical properties of the mobile phase. Examples are the differential refractometer, conductivity and dielectric constant detectors. The general principles of some of these detectors are described; however, special emphasis is placed on the more common modes of detection, i.e., UV and RI.

The ultra-violet detector records the adsorbance of a component or molecule at a fixed, specific wavelength. The more commonly fixed wavelength detectors availavle are 254 and 280 nanometers (nm). This detector offers great sensitivity, i.e., 0.0001 adsorbance units and a linear dynamic range of approximately 5×10^3.

The source of energy for these UV detectors consists of a low pressure, hot cathode mercury lamp. Because of the penetrating power of UV radiation, a wor d of caution. . . . . .with the naked eye, do not look directly into the cell or source lamp while in

operation! One should always wear safety glasses or
other eyeglasses.

The UV detector is not universal and must be accepted
as such because many materials do not exhibit UV
absorption. This limits its utility to compounds
which adsorb UV radiation. Byrne (1) stated that
non-UV adsorbing compounds may produce signals in the
UV detector as the result of changes in refractive
index due to light scattering phenomena.

There are numerous suppliers of UV detectors and all
function on basically the same principle with some
minor differences. The principle of operation of
the UV detector consists of light sensitive resistors
arranged in a Wheatstone bridge circuit and the
sensitivity of these resistors is universely proportional to the UV light impinging upon them. When
a UV adsorbing component enters the cell from the
column an imbalance is recorded as a peak on the
chromatogram.

Figure 5-1 shows a diagram of the Waters Model 77 UV
detector. Two beams of radiation from a common area
on the source lamp, S, pass through lens, L, flow
cell chamber, C1 and C2, plane window, W, a visible
light blocking filter, F, and finally impinge upon
dual photodetector sensitive areas, D1 and D2. The
radiation passing through the cell chambers is collimated. As the cell chamber axes are mutually
parallel, the result is that the two beams passing
through the chambers have originated from virtually
the same area on the source lamp, S. The use of a
common source area for the two beams is advantageous
because spatial variations in lamp brightness as
a result of temperature changes, dust accumulation,
droplets of mercury condensate, etc., are "common
mode" to both beams and do not effect the detector
readout.

In addition to its low cell volume (10 μl) which
does not degrade chromatographic separations, the
system is inherently stable due to a single optical
axis and a single set of optics common to both
beams.

Most UV detectors have a meter displaying the bridge
excitation voltage. With the bridge voltage

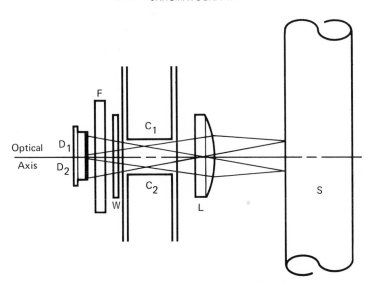

Figure 5-1 Optical Path of a Typical UV Detector
(Courtesy of Waters Associates)

balanced this meter should read within certain limits
as described in the manual for that detector. The
function of this meter is to assure the user that
sufficient radiation is reaching the photodetector
elements. The output of this meter will increase
in the presence of large signals. This meter is an
extremely valuable trouble-shooting device. Air
bubbles passing through the detector may be diagnosed
quite readily merely by observing the meter output.

An air bubble build-up may also be observed. A
higher than normal meter output indicates that the
eluant (solvent) contains possibly a small amount
of adsorbing material(s), perhaps added to the solvent by the supplier to function as an inhibitor.
A classic example is that of the solvent tetrahydrofuran (THF) containing small amounts of the inhibitor ionol.

The bridge excitation meter may also signal the user
when cleaning of the detector is needed. Again,
higher than normal output on the meter may indicate
component film deposits within the cell, which may
have occurred during previous runs with different
solvents and solutes. The recommended approach is
that of cleaning with a miscible solvent series.

A choice of two detection wavelengths is generally
available for most UV detectors. 254 nm is the more
common choice; however, there are numerous compounds
which exhibit greater sensitivity at 280 nm. This
choice is provided by UV filters which are easily
changed. In humid or dusty areas these filters may
become coated either with condensate or dust resulting in erratic recorder response or a decrease
in sensitivity. A periodic check of these filters,
subsequently wiping with a non-abrasive, lint-free
cloth or napkin will eliminate these problems.

A general rule of thumb regarding the detection
limits of UV detectors at 254 nm is seen in Table
5-1. For example, at 254 nm, the detectability of
the nucleic acid base, uracil, is approximately
5×10^{-10} gm./ml.

The differential refractometer detector is considered
by many to be the "universal" detector. This is not
necessarily true, as all of the current LC detection

Table 5-1

General Capability of Detection Limits
for UV Detectors

Extinction Coefficient (ϵ)	Detection Limit (gms.)
10^4	10^{-9}
10^3	10^{-8}
10^2	10^{-7}
10	10^{-6}

systems have limitations. In defining a universal
detection system for LC, one must specify a type of
detector that is sensitive to all classes and types
of compounds, and perform this admirably when subjected to wide variations in operating parameters
such as temperature, flow rates, pump (pressure)
pulsations, and solvents.

The refractometer does provide a wide range of application in that all substances have a refractive
index. Sensitivity is approximately 10^3 lower than
the UV detector; however, the RI detector possesses
the advantage of exhibiting "universal" sensitivity
provided the RI of the solute and solvent are different.

The major disadvantages of the RI detector are
those of extreme sensitivity to temperature changes,
flow rate fluctuations and lack of applicability for
gradient elution work. Column ovens (Chapter 4)
and pulseless pumps (Chapter 2) assist in minimizing
the effects of temperature and flow variations.

The two designs of refractometers available consist
of the Fresnel type and the deflection type.

Figure 5-2 is a schematic diagram of the Fresnel
type refractometer detector patented by E.S. Watson
(2). This refractometer uses a single-axis optical

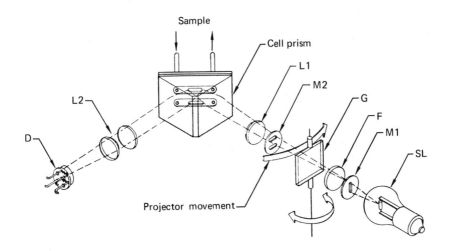

Figure 5-2   Schematic Diagram of the Fresnel Type
             Refractometer

system with common optics and a single prism to il-
luminate the two interfaces. Watson states that
inherent stability is obtained in addition to reducing problems associated with temperature fluctuations. Transmittance rather than reflectance of
the interfaces is measured.

In reference to Figure 5-2, light from source lamp
SL passes through source mask M1, infrared blocking filter F, fine adjusting glass G, aperture mask
M2, and is collimated by lens, L1. Mask M2 defines
two collimated beams that enter the cell prism
and impinge upon the two glass-liquid interfaces
formed by the sample and reference liquids, which
are in contact with the prism. SL and L1 are
mounted in a common assembly called the projector
which can be rotated about the axis of the prism.
This permits coarse adjustment of the incident angle
to slightly less than the critical angle and is made by
rotating the fine adjusting glass, G. Light from
the two beams which is internally reflected does not
enter detector lens L2, only that light which is
transmitted through the two interfaces passes through
and impinges on the stainless steel plate. The surface of this plate has a finely ground light scattering surface and the transmitted beams appear as
two spots of light. The detector lens assembly forms
an image of these spots on two light sensitive
elements in the photodetector, D. The photoconductor
elements in the dual photodetector are arranged in
a Wheatstone bridge circuit whose imbalance provides
the RI measurements.

Short term peak to noise is equivalent to 3×10^{-7}
RI units. Drift rate is less than 1×10^{-6} RI
units per hour. Another advantage of the Fresnel
detector is that only a very thin film of liquid is
required for a measurement.

Fresnel equations are non linear for parallel mono-
chromatic rays; however, adequate linearity is
obtained in the region of 10% transmittance.

Disadvantages of the Fresnel refractometer are that
unstable films may form in the prism and affect
measurements; additionally, these refractometers are
more sensitive than deflection refractometers to
bubbles and particulate matter in the cells.

Figure 5-3 is a schematic diagram of the Waters
Associates deflection refractometer detector. This
detector measures the deflection of a light beam
resulting from the difference in RI between the sample and reference liquids. Deflection occurs at
the surface of the cell partition permitting the use
of small cell volumes. In principle, a beam of light
from the lamp passes through the optical mask which
confines the beam to the region of the sample cell.
The mirror reflects the beam back through the sample
and reference cells and through the lens coming into focus on the detector. The angle of deflection
(RI between sample and reference) is determined by
the location of the focused beam on the detector.
As the beam moves on the detector, an output signal
is generated, amplified and recorded. The optical
zero glass deflects the beam from side to side to
adjust for zero output signal. Advantages of

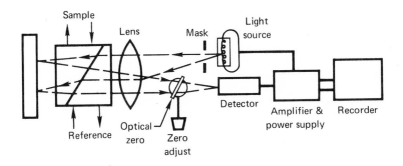

Figure 5-3   Deflection Refractometer Detector
             (Courtesy of Waters Associates)

deflected (transmitted and refracted) beam detection
consist of wide range solvent capability, minimal
signal interference by bubbles or particulate matter
in the solvent, and the inherently high sensitivity
obtained.

In addition to the inherent disadvantages of refractometer detectors in general, such as drift caused by
temperature changes and erratic responses due to
pressure fluctuations, the deflection refractometer
is not as cleanly swept as the Fresnel detector

when used for extremely high speed applications. When all parameters are optimized changes as small as 10^{-7} RI units are claimed to be detectable.

The solute transport detector or flame ionization detector described by Gilding (3) involves the removal of the mobile phase prior to detection. In principle, the column eluant containing the solute to be analyzed is fed onto a moving wire or conveyor. The mobile phase is evaporated leaving the solute deposited on the wire which is subjected to a pyrolysis chamber containing a special catalyst for producing a smooth decomposition reaction. Advantages of this detector are that the mobile phase has no effect on the detector, linear response is obtained, and responds to virtually any compound containing carbon. Disadvantages are also evident in that the system is limited to non-volatile solutes, comparatively poor sensitivity is obtained and it is quite bulky and expensive.

The maintenance and trouble-shooting regions of major concern in the solute transport detector are those of complete mobile phase evaporation, activity of the catalyst bed and reproducibility.

The micro-adsorption detector, M.A.D., (Varian Aerograph) (4), consists of two cell compartments; a reference cell packed with non-adsorbing material and an active cell packed with an adsorbent. A thermistor probe is mounted in each cell. When a component from a mixture is adsorbed the thermistor temperature increases due to the heat of adsorption. Upon desorption, the thermistor temperature decreases. The measure of these temperature changes comprises the principle of the M.A.D. A Wheatstone bridge is used for measuring the temperature imbalance between the sample and reference cells.

The signal of an ideal micro-adsorption detector would be the differential of a gaussian curve. However, in practice, the signal departs from this ideal shape because of heat loss to the eluant stream and the environment. Advantages of the micro-adsorption detector are its minimal dead-volume and simplicity. Disadvantages are those of precise temperature control as the sensitivity is an inverse function of the detector temperature. One may experience a twenty to eightly percent loss in sensitivity for a

35°C change in temperature. The activity of the mobile phase influences sensitivity also as it competes with the solute for the adsorbent packing.

Fluorimeters for liquid chromatographic detection are becoming more widely used. Detection is based upon the fluorescent energy emitted from a solute excited by UV radiation. Fluorescent detectors are generally no more sensitive than conventional UV detectors; however, they do provide high selectivity. The major disadvantage associated with fluorescence detectors is its susceptibility to interference by fluorescence or quenching effects from background.

LC detectors have many sources of noise. The four major causes of noise are those associated with the electro-optical system, temperature fluctuations, fluctuations in chemical composition and flow rate and pressure fluctuations.

High frequency noise is defined as greater than one cycle per five seconds. This noise generally has no effect on the detection limits because of its high frequency and low amplitude. Short term noise is within the range of one cycle per five seconds to one cycle per five minutes. It is the limiting noise for fast and medium speed peaks. Drift is a baseline shift over a period greater than five minutes. The major factor contributing to this drift is ambient temperature change. This noise does not affect the detection limit of slow eluting components.

Temperature fluctuations, as mentioned earlier, cause baseline drift; however, with UV detectors this is minimal compared to differential refractometers. If the detector is housed in a well ventilated compartment (fan, vents, etc.) the problem is minimized.

Fluctuations resulting from variations in the chemical composition of UV adsorbers in the solvents being used will produce large baseline irregularities. Table 5-2 (5) shows the relative UV absorbance of common solvents listed in an eluotropic order. The length of the horizontal line from each solvent depicts the regions it cannot be used in UV detection.

TABLE 5-2

RELATIVE UV ABSORBANCE OF SOLVENTS (5)

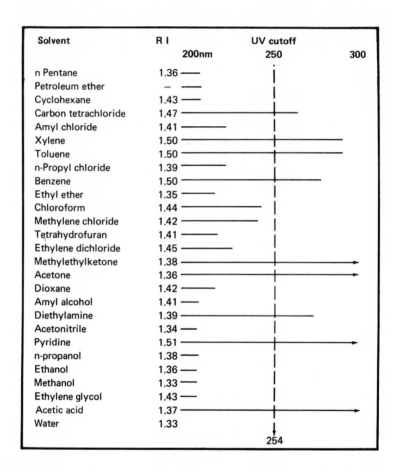

Solvent	R I	UV cutoff		
	200nm		250	300
n Pentane	1.36			
Petroleum ether	–			
Cyclohexane	1.43			
Carbon tetrachloride	1.47			
Amyl chloride	1.41			
Xylene	1.50			
Toluene	1.50			
n-Propyl chloride	1.39			
Benzene	1.50			
Ethyl ether	1.35			
Chloroform	1.44			
Methylene chloride	1.42			
Tetrahydrofuran	1.41			
Ethylene dichloride	1.45			
Methylethylketone	1.38			
Acetone	1.36			
Dioxane	1.42			
Amyl alcohol	1.41			
Diethylamine	1.39			
Acetonitrile	1.34			
Pyridine	1.51			
n-propanol	1.38			
Ethanol	1.36			
Methanol	1.33			
Ethylene glycol	1.43			
Acetic acid	1.37			
Water	1.33			

254

The effects of flow rate and pressure fluctuations upon UV detectors is minimal; however, some do exist and result in noise. In order to obtain the maximum UV sensitivity in high pressure systems, it is preferable to by-pass the reference cell because even flow rate changes are translated into pressure changes in the reference cell. The use of pulseless pumps (see Chapter 2) eliminates problems associated with pressure pulses.

Bubbles resulting from improper system purging, leaks, and solvents which have not been degassed, present the user with unnecessary problems and contribute to appreciable down time. As mentioned earlier, these bubbles may be observed either on the bridge excitation meter or the recorder trace upon passing through the detector. Corrective measures are pre-purging the column, tightening and/or replacement of leaking fittings or septa, and proper degassing of solvents. Problems may occur, however, when attempting to dislodge an entrapped bubble in any "larger-than-necessary" dead-volume area. This occurs quite frequently in aqueous systems and in high vapor pressure organic systems.

Several techniques may be used depending upon location of the bubble-trouble. Some workers have found that placing considerable back-pressure on the system will displace the lodged bubbles. This is accomplished by fitting a plug (septum or silicone rubber) onto the end of the detector exit or exit tube during operation. The subsequent pressure build-up will often free the system of bubbles. This technique must be used with care as the detector may become damaged. Increasing the pressure delivered by the pump will often be sufficient to dislodge the entrapped bubble, especially in the vicinity of the injector and pre-column plumbing. However, most problems seem to occur in the vicinity of the detector, either within or immediately before entering the detector. First of all, the exit line from the detector to the waste resevoir should be disconnected. The meter and/or the recorder trace should then be observed. This exit tube often becomes partially plugged placing back-pressure on the detector as in gas chromatography. This tube may either be cleaned with miscible solvents or replaced. If this does not correct the situation, back-flushing of the detector and associated transfer lines and fittings from the column exit are in order.

During normal operation, the pressure-drop along the column is quite large resulting in only minimal pressure being imposed at the entrance to the detector. Entrapped bubbles may be successfully purged-out of the system by circumventing the column and adapting flow directly from the pump. CAUTION....

When using this procedure, keep the pump pressure at a minimum lest the cell be damaged. Ten ml. hypodermic syringes may alternatively be used rather than the pump for purging the cell.

References

1. S.H. Byrne, Jr. in Modern Practice of Liquid Chromatography, J.J. Kirkland, Ed., John Wiley and Sons, Inc., New York, 1971.

2. E.S. Watson, American Laboratory, September 1969.

3. D.K. Gilding, American Laboratory, October 1969.

4. F.R. MacDonald, C.A. Burtis, and Jack M. Gill, Research Notes, Varian Aerograph, July 1969.

5. LAB NOTES, 4, CHROMATRONIX Incorporated, January 1971.

Chapter 6

COMPREHENSIVE TROUBLE-SHOOTING OF LC SYSTEMS

As most analytical problems in liquid chromatography
are approached using the "sequential analysis" or
"first things first" method, virtually the same
approach may be employed in developing a comprehensive system for trouble-shooting of the LC equipment.
In liquid chromatography possible "trouble areas"
are quite numerous and may be associated with any
one major part of the system or a multiplicity of
several malfunctions, in combination, from more than
one major part of the system. Major parts of the
system are: the solvent transport system, sample
introduction, column, detector and recorder.

A comprehensive trouble-shooting guide should provide a method for rapidly pinpointing the symptoms
and causes associated with a particular malfunction.
Additionally, it should provide guideline remedies
and corrective procedures for restoring the equipment to its maximum capability.

Management, upon spending thousands of dollars for
equipment and associated apparatus, needs the as-
surance that its investments are being utilized to
the utmost. This places an added and well-justified
responsibility upon all who design, sell, purchase,
and use liquid chromatographic equipment. Recogn-
izing that no system, chromatographic or otherwise,
is exempt from maintenance and trouble-shooting
problems, it behooves us as buyers and users, to
make wise choices in our purchases and also maintain the optimum in performance from our equipment.

Preventive maintenance is a milestone in achieving
maximum performance. A properly executed preventive
maintenance program will save costly downtime. A
preventive maintenance program cannot provide nor
guarantee trouble-free operation. Therefore, when
anomalies inadvertently occur, we should be prepared
to rapidly diagnose the problem and either remedy or

provide guidance to the subsequent restoration of
the equipment to its optimum capability.

The ultimate solution for accomplishing this is
either via training and experience or a readily ac-
cessible guide which may partially substitute for
the lack of training or experience. The following
is such a guide which may provide rapid troubleshooting tips for the experienced chromatographer as
well as the novice in the field. The general format
of the guide will be a description of sympto ,
causes, and corrective actions for each major category of the liquid chromatographic system. These
are:

I. Solvent Transport System
Solvent degassers, solvent reservoirs, solvent
pumps, and associated transfer lines.

II. Sample Injection System
Septum injectors and loop injectors. Solvent
leaching of septa, leaking of septa and loop
injectors,   particulate matter.

III. Columns
Column bleed, leaking fittings, column plugging.

IV. Detectors
Low sensitivity, non-linearity, noise, erratic
response.

V. Recorders
Malfunctions diagnosed from recorder traces,
recorder malfunctions, noise, electrical connections.

In using the refractometer, flow trouble-shooting
may become important. It is often difficult to determine if the flow stream or the refractometer is
causing the problem. To solve some of these
problems the following procedure may be used:

1. Stop the flow. Normally, the baseline read-
out will be different. Criteria such as
drift due to mixing, fractionation, and
cycling will virtually be eliminated. If
anomalies do occur then one must suspect the
detector.

2. Dual-column, matched-flow systems are recommended for determining anomalies in differential refractometer applications.

3 Non-linearity is often the problem in refractometer monitoring. Virtually the only
   cause for this is misalignment of the image
   transmitted to the photocell. This indicates
   that the beam is not properly focused or,
   the beam does not pass through the cell
   properly. This may occur with either type
   of refractometer, Fresnel or Deflection.
   Proper alignment is mandatory if anything
   other than qualitative determinations are
   being sought. Quantitatively, accuracy of
   the technique should be at least $\pm 5\%$,
   relative to the amount present for any given
   component.

Concluding, the trouble-shooting tips outlined in
this chapter were obtained from several sources (1,
2,3) as well as our own experiences and are representative of most malfunctions which may occur in liquid
chromatography.

Symptom	Cause	Corrective Action
Noisy Baseline (or)	**A. Detector** 1. Contamination in sample or reference cells or bubbles in sample cells or reference cell. 2. Bubbles in sample cells or reference cell. 3. Defective UV lamp.	**A. Detector** 1. Flush sample and reference cells with fresh solvent. 2. Increase flow rate or place restriction at detector exit. 3. Replace UV lamp.
	B. Solvent Transport System 1. Bubbles passing through sample or reference cells. Solvents not adequately degassed. 2. Pulses from pump stroke.	**B. Solvent Transport System** 1. Degas solvent. 2. Incorporate pulse dampener or reduce pump stroke.
	C. Column 1. Particulate matter from column passing through or into detector. 2. Leaking fitting or connector.	**C. Column** 1. Check column exit. 2. Tighten or replace fittings.
	D. Sample Introduction System 1. Leaking septum. 2. Partial blockage in loop injector due to particulate matter.	**D. Sample Introduction System** 1. Replace septum. 2. Clean loop injector.
	E. Recorder 1. Grounding problem in recorder or instrument.	**E. Recorder** 1. Check all recorder and instrument grounds. Eliminate ground-loops.

58

Symptom	Cause	Corrective Action
Drifting Baseline (short-term and long-term drift)	A. Detector 1. Contamination in sample or reference cells. 2. Changes in temperature of detector. 3. Contamination build-up in cell. 4. Dust or condensate contamination in optical system. 5. Weak source lamp. B. Solvent Transport System 1. Bubbles in mobile phase. 2. Contaminant in solvent resevoirs slowly being dissolved into mobile phase. 3. Previous solvent not completely removed. 4. Solvent demixing (or) nonhomogeneity. C. Column 1. Column bleed.	A. Detector 1. Flush cells with solvent. 2. Control temperature using constant temperature bath or dual-column arrangement. 3. Clean cell using 6N nitric acid and distilled water. 4. Wipe clean with lint-free tissue or cloth. 5. Replace source lamp. B. Solvent Transport System 1. Degas mobile phase. 2. Wash resevoirs and replace old solvent with freshly prepared one. 3. Allow adequate purge-time. 4. Insure solvent compatability. C. Column 1. Saturate mobile phase with stationary phase or use a pre-column.

59

Symptom
Drifting
Baseline

(short-
term and
long-term
drift)

Cause
C. Column (con't)
22. Strongly adsorbed
component(s) being
eluted from col-
umn.

D. Sample Introduction System
1. Septum leaching caused
by septum/mobile phase
incompatability.
2. Plugging of injector
by particulate matter
from solvents or septa
causing pressure
changes.

E. Recorder
1. Weak tubes in
amplifier.

Corrective Action
C. Column (con't)
2. Elute components from column. Use elustropic solvent
series when necessary.
Select mobile phase with
greater compatability for
all components of mixture,
or increase temperature.

D. Sample Introduction System
1. Select recommended septum
for use with various mobile
phases.
2. Clean sample introduction
system.

E. Recorder
1. Check tubes and replace if
necessary.

60

Symptom

Baseline "Stairstepping" and peaks are "Flat-topped." Baseline does not return to zero.

Cause

A. Recorder
   1. Gain and damping control not properly adjusted.
   2. Recorder or instrument not properly grounded.

Corrective Action

A. Recorder
   1. Adjust gain and damping controls.
   2. Properly ground via a true earth ground.

Symptom

"Spiking" i on recorder trace

Cause

A. Detector and Solvent Transport
   1. Bubbles passing through detector (or) bubbled lodged at detector entrance.

B. Other Systems
   1. Externally located electrical systems "tapped" off a common electrical supply, i.e., other chromatographs, thermostated oven, etc. causing intermittent line-voltage flucuations feeding-back to the LC system.

Corrective Action

A. Detector and Solvent Transport System
   1. Degas eluants and purge entire system adequately.

B. Other Systems
   1. Eliminate possible line voltage feedback fluctuations from other systems. Check spikes for time regularity and magnitude (duration).

61

Symptom
Negative peaks on recorder trace

Cause
A. Detector
1. Detector output polarity.
2. In using RI detectors, some components in the mixture may have RI greater or less than mobile phase.
3. Sometimes observed when using UV detectors and non-UV adsorbers pass through detector. It is actually a RI measurement due to light-scattering phenomena.

B. Solvent Transport System
1. Large quantities of air in mobile phase.

C. Column
1. Air pocket displaced from column.

D. Sample Introduction System
1. Negative peak upon injecting sample.
2. Septum begins leaking upon injection.
3. Injection of sufficient quantities of air.

Corrective Action
A. Detector
1. Change polarity, + or -, of detector output switch.
2. Select mobile phase of greater or less RI, (or) set detector and recorder "zero" at mid-scale initially.
3. No corrective action.

B. Solvent Transport System
1. Degas mobile phase.

C. Column
1. Adequately purge column with mobile phase.

D. Sample Introduction System
1. Install loop injector.
2. Replace septum.
3. Displace air from syringe before injection. Purge sample loops free of air before introducing into the system.

Symptom
Poor Peak
Shape

Cause
A. Column
1. Column Overload.

2. Adsorption of
   sample on ion exchange
   (IEC) column.

Corrective Action
A. Column
1. Reduce amount of sample. For
   GPC, use apprx. 0.25% solution of sample/solvent. For
   LLC and LSC, use apprx. 0.05
   to 0.10% solution of sample/
   solvent. For IEC using pellicular ion exchange resins,
   anything in excess of 50 to
   100 ug of any component may
   result in column overload.

2. Lower sample concentration or
   increase the ionic strength
   of the eluant. Altering the
   pH of the eluant will have
   marked effects also. Increasing the column temperature will subsequently increase the ion exchange
   process.

Symptom
Loss of
Resolution

Cause
A. Column
1. Loss in column ef-
   ficiency.
2. In LLC, loss of liquid
   phase from folumn.

Corrective Action
A. Column
1. Replace column or regenerate
   column (IEC).
2. Replace column and use a mobile
   phase that does not remove
   liquid phase. Also, permanently bonded liquid phases
   should be considered, i.e.,
   Permaphase, (DuPont) whereby

Symptom
Loss of
Resolution

Cause
A. Column (con't)
  2. In LLC, loss
    of liquid phase
    from column.
  3. Column Overload.
  4. Increase in column
    flow rate.

Corrective Action
A. Column (con't)
  2. removal of liquid phase by
    the mobile phase cannot occur.

  3. Lower sample concentration.
  4. Decrease flow rate.

Symptom	Cause	Corrective Action
Increased Retention Volumes	A. Column 1. In LSC, activity of column is increasing. Solvent is stripping water from column. 2. In LLC, liquid phase has been lost from the column. 3. Temperature of column is too low.	A. Column 1. Add water to mobile phase or replace with new column which has been properly deactivated. 2. Replace column and change mobile phase. 3. Increase column temperature.
	B. Solvent Transport System 1. Flow rate of mobile phase is too low. 2. In IEC, ionic strength of mobile phase is too low. 3. In IEC, pH may be too high or too low.	B. Solvent Transport System 1. Increase flow rate. 2. Increase ionic strength of mobile phase to shorten retention volume. 3. Alter pH; this may alter the elution order also.

Symptom
Recorder
will not
"zero"

Cause
A. Recorder
1. Power to pen not
   turned "ON"

2. Recorder putput
   "dead"

B. Detector
1. Poorly connected
   electrical leads from
   detector output to
   recorder terminals.

2. Bubble in sample cell
   or reference cell.

3. Physical blocking of
   sample or reference
   image from detector.

4. Source lamp defective

Corrective Action
A. Recorder
1. Check power switch. If already turned "ON", switch
   may be defective.

2. Check recorder "ZERO" by
   shorting the input from the
   detector, i.e., turn attenuator switch to infinity
   (∞) and adjust recorder
   zero. If this is possible,
   problem is probably associated with the input signal from
   the detector. If, in the
   position, the recorder will
   not zero the problem is associated with the recorder.

B. Detector
1. Check and secure electrical
   leads from detector to recorder.

2. Increase flow rate to purge-
   out bubble or place restrictor at detector exit
   to displace bubbles.

3. Check light path and remove
   obstructions.

4. Replace source lamp.

66

Symptom
Recorder
will not
"zero"

(con't)

Cause
C. Column
1. Excessive column
   "bleed".
2. Air being displaced
   from column.
3. Previous mobile phase
   not removed.

D. Solvent Transport System
1. Previous solvent
   not removed.

Corrective Action
C. Column
1. Use balancing dual-column
   technique.
2. Adequately purge column
   with mobile phase.
3. Purge column free of previously used eluants.

D. Solvent Transport System
1. Allow pump and associated
   transfer lines to be completely purged-free of
   previous eluants.

67

Symptom
No flow through column and no pump pressure

Cause
A. Solvent Transport System
1. Solvent resevoir empty.
2. Air in pump.

Corrective Action
A. Solvent Transport System
1. Refill resevoir with desired solvent.
2. Siphon liquid into pump and resume pump action.

Symptom
Pump exerting pressure on system, but no flow through column

Cause
A. Solvent Transport System
1. Leak in system.

2. Restriction to flow in the vicinity of the pump and the column exit to the detector.

B. Sample Introduction System
1. Plugging in the sample in inlet due to particulate matter from septa, samples, etc.

2. Syringes

Corrective Action
A. Solvent Transport System
1. Check septum, relief valve (defective?), transfer line from pump to injector. Check all associated fittings and couplings from pump to detector exit.

2. Particulate matter will clog the small diameter lines to the pulse dampners on pulsating pumps; therefore, isolate these before encountering downstream pursuits.

B. Sample Introduction System
1. Filter all solvents, even distilled H_2O.

2. Exercise care with syringes. Use syringes that are free of barbs, etc., which may tear rather than pierce the septum upon sample injection.

Symptom
Pump exerting pressure on system, but no flow through column.

Cause
C. Column
1. Leaks in fittings.

D. Detector
1. Leaks.

Corrective Action
C. Column
1. Check all fittings for leaks.

D. Detector
1. Check detector and associated fittings.

69

References

1. N. Hadden and F. Zamaroni, Troubleshooting and Maintenance, Varian Associates, 1970.

2. L.R. Snyder and J.J. Kirkland, Modern Liquid Chromatography (ACS Short Course), American Chemical Society, 1971.

3. Instruction Manual (98526), Waters Associates, Inc., 1970.

GAS CHROMATOGRAPHY

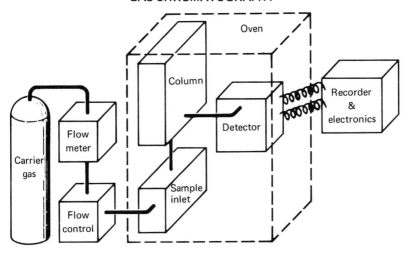

PART II. GAS CHROMATOGRAPHY

This part of the text is concerned with Gas Chromatography (GC) and begins with the basic fundamentals. Following this brief introduction chapter, there are practical approaches to understanding and maintaining the following GC systems: the Carrier Gas-Inlet, the Column-Oven, the Detector, the Recorder and Electronics. A final chapter on Comprehensive Trouble Shooting, is a ready reference for fast GC problem solving.

One section of Part II includes maintenance and problems with chromatographic integral electronics (i.e., power supplies, electrometers, recorders, and integrators). Another section includes maintenance of auxillary equipment interfaces (i.e., ultraviolet, infrared as mass spectrometers). A trouble-shooting section concerning thermal degradation (pyrolysis) inlet systems is also included in Part II.

CHAPTER 7

INTRODUCTION TO GAS CHROMATOGRAPHY

Gas chromatography (GC) is an instrumental method of analysis for the separation, identification, and quantitation of volatile mixtures. "Permanent" gases (such as oxygen and carbon dioxide), volatile liquids, and pyrolyzed solids can all be separated by gas chromatographic techniques. Part II of the text will be concerned primarily with the GC instrumentation which permits these analysis to be carried out. Some basic knowledge of the principles of separation is necessary, however, for a full appreciation of the material presented in subsequent chapters. A detailed explanation of the theory will not be necessary. For a more rigorous development of the theory, texts by Keulemans (1), Zlatkis (2), and others should be consulted.

This first chapter, dealing with the basic fundamentals of gas chromatography, includes some important definitions and formulas which are used to describe the performance of GC instruments. In later chapters the characteristics, the maintenance, and the troubleshooting of various segments of the gas chromatograph will be discussed separately including inlet systems, columns, detectors, and electronics. Finally, a summary chapter is included which is a comprehensive guide for determining the causes and remedies for problems occuring in the GC.

The Principle

Separation is the primary function of a gas chromatograph. This process of separation takes place inside the column; a length of tubing containing a packing material which consists of a stationary phase coated on an inert support. As illustrated in Figure 7-1, the mixture, consisting of components A and B, is introduced into a gaseous mobile phase which is constantly traversing the length of the column. While moving through the column, components A and B will interact with the stationary phase

While the mobile phase does not interact at all.
For this reason, A and B are delayed as they move
through the column. In Figure 7-1, component B in
teracts with the stationary phase to a greater extent than component A and, consequently, the mixture
begins to separate. Ideally, the mixture is completely separated by the time the components reach
the end of the column.

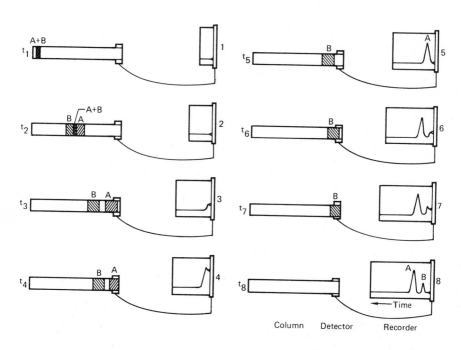

<u>Figure 7-1</u>. The operation of a chromatograph is
shown in its most basic concept. A mixture of components A and B enters the separation column, at
left, where it is separated into pure A and pure B.
The arrival of the components at the end of the column
is observed by the detector which signals a strip
chart recorder. The end result is a cromatogram
which is representative of the original mixture.
(Courtesy "Chemistry" and American Chemical Society)(3)

The elution of the components from the column is detected electronically by a suitable detector, and
signals are sent to a strip chart recorder which
produces the chromatogram. This chromatogram consists of a series of peaks, each of which indicates
the elution of a component and the amount of component present. The time of elution may be used to
identify the components of the mixture and is defined as the elution time or retention time for a
particular component.

A basic gas chromatograph, consisting of the following systems is shown in Figure 7-2:

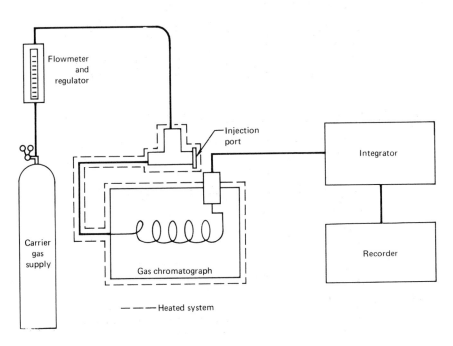

Figure 7-2. The basic parts of a GC system are shown
with the heated areas surrounded by broken lines. An
integrator is sometimes employed to measure the area
under the peaks of the chromatogram. (4)

I. Pneumatic and Sample Systems (Chapter 8)
    A. Carrier Gas (mobile phase)-Inert gas used
        to move the sample through the column.
    B. Pressure and/or Flow Control Apparatus-
        Maintains a constant pressure and/or flow
        rate of carrier gas through the column. It
        may also vary column pressure at a prede-
        termined rate.
    C. Sample Port (injector)-for introduction
        and vaporization of the sample.
    D. Pyrolyzers-for the thermal degradation of
        solid and liquid samples prior to intro-
        duction to the column. (Chapter 9).

II. Columns and Column Ovens (Chapter 10).
    A. Separating Column-tubing containing the
        stationary phase which is coated either on
        an inert solid support or on column walls.
    B. Ovens, Heaters, and Controllers-controls
        the temperature of the column, detector,
        and injector.

III. Detectors and Related Electronics (Chapter 11)

    A. Detector-Detects sample components as they
        elute from the column and provides the basis
        for quantitative measurement. (Chapter 11).
    B. Power Supply and/or Electrometer Cir-
        cuits-Amplifies the detector signal
        which is sent to the recorder and pro-
        vides the detector with needed power
        (Chapter 12).

    C. Recorder-Provides a permanent visual record
        of the analysis (Chapter 12).

IV. Auxillary Systems (Chapter 13).
    A. Mass Spectrometers, infrared spectrophoto-
        meters, and ultra-violet spectrophotometers
        aid in the positive identification of the
        sample.
    B. Mechanical and digital integrators aid in
        quantitating chromatographic peaks.(Chapter 12).

In an ideal situation, the peaks of the chromatogram-should be completely separated. The peak width at the base should be narrow, even at the end of the chromatogram. These performance factors are measured in terms of _efficiency_ and _resolution_ which can be calculated from a few simple measurements made on the chromatogram. A typical chromatogram is shown in Figure 7-3. The time required for the peak to appear, called the retention time (t_R), is measured from the point of sample injection to the point at which the peak height is at its maximum. The width of the peak (W) is measured at the bottom of the peak (base line) between lines drawn tangent to the sides of the peak shown in Figure 7-3.

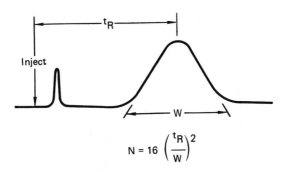

$$N = 16 \left(\frac{t_R}{W}\right)^2$$

Figure 7-3. A chromatogram consists of a set of peaks produced by the chart recorder. The retention time of a peak (t_R) is measured from the point of injection to the highest point of the peak. Peak width (w) is measured at the base of the peak between lines drawn tangent to the sides of the peak. From these two dimensions, column efficiency (N) can be calculated. (Courtesy Varian Aeograph) (5)

Column efficiency, given by the _number of theoretical plates_ (n) is calculated from the expression,

$$N = 16\frac{t_R^2}{w_b}\qquad \text{(approximate value)}$$

where,

t_R = retention time (here measured in units of length).

w_b = peak width at the base.

N = number of theoretical plates.

Efficiency is a function of column length among other parameters; and, for purposes of comparison, efficiency is commonly expressed in terms of <u>Height Equivalent to a Theoretical Plate</u> (HETP). HETP is the column length (L) divided by the number of theoretical plates (N), as follows:

$$HETP = \frac{L}{N}$$

A third expression for column efficiency is the reciprocal of HETP which expresses efficiency in terms of plates per foot. All three expressions for efficiency are used.

The ability of a column to separate a given pair of compounds can be measured in terms of resolution (R). As illustrated in Figure 7-4, resolution is a function of peak width and distance between peaks.

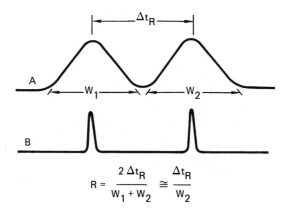

$$R = \frac{2\,\Delta t_R}{W_1 + W_2} \cong \frac{\Delta t_R}{W_2}$$

<u>Figure 7-4</u>. The ability of a particulat column to separate a given pair of compounds can be measured in terms of resolution (R). As illustrated, R is a function of peak width and the distance between two peaks. Usually R is measured for the pair of peaks which are the most difficult to separate. (Courtesy Varian Aerograph) (5)

R = column resolution
W_1, W_2 = peak width at base
t_R = distance between the apex of the two peaks

A very useful equation for comparing column dimensions and analysis conditions was proposed in 1956 by Van Deemter (3). The simplest form of the equation, relating HETP and carrier gas velocity, is written as follows:

$$HETP = A + \frac{B}{\bar{\mu}} + C\bar{\mu}$$

This equation can be plotted with HETP as a function of carrier gas velocity ($\bar{\mu}$) as shown in Figure 7-5. The curve has a minumum at a particular value of $\bar{\mu}$ called the optimum carrier gas velocity ($\bar{\mu}_{opt}$). At this point HETP is at it's lowest value and the column efficiency is, thus, at its highest.

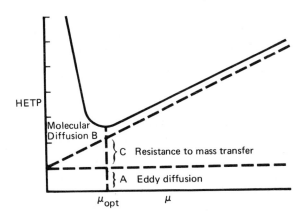

Figure 7-5. Van Deemter proposed an equation relating HETP to carrier gas velocity. A, B, and C are constants which describe the physical characteristics of the column affecting HETP. A plot of HETP versus carrier gas velocity will reveal a curve with a minimum at a particular carrier gas velocity which should provide the highest column efficiency. (6)

Van Deemter plots can also be used to compare various types of columns and stationary phases (6). The effect of HETP on the amount of liquid stationary phase used in gas-liquid chromatography is shown in Figure 7-6. The percent liquid phase is a relative figure representing the percentage of liquid phase in the solution used to coat the column or column support. Columns with a liower liquid load exhibit lower minumums of HETP and are thus capable of higher efficiency.

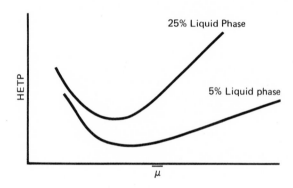

Figure 7-6. Lower amounts of liquid phase (determined by the percentage of liquid phase in the solution which was used to coat the column support) will provide more efficient columns. (5)

HETP is dependent upon the type of carrier gas used. Some carrier gasses are able to diffuse more readily through the solid support. They are, as a result, better able to move through the column. Relative van Deemter plots for nitrogen and helium carrier gas are shown in Figure 7-7. Nitrogen, having a higher diffusivity, displays a lower minimum HETP.

Column diameter affects the van Deemter plot as shown in Figure 7-8. Reducing the internal diameter of a column, all other factors being held constant, will result in a more efficient column. Difficulties in packing small diameter columns with stationary phase places a lower limit on the size of column that can be used, but columns with internal diameters as low

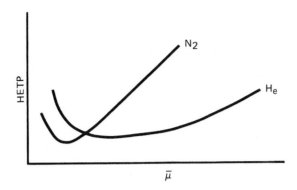

Figure 7-7. Different carrier gases exhibit different
HETP plots. Helium carrier gas has a wider range of
good carrier gas flows, however nitrogen is capable
of higher efficiency. (5)

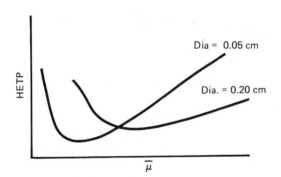

Figure 7-8. Smaller diameter columns are more efficient but require smaller samples. The minimum
practical column diameter (due to packing difficulties)
is about 0.03 cm. (5)

as .025 cm are used frequently in situations where
high resolution is required. The small diameter
columns also require the use of comparitively smaller
sample sizes.

If too much sample is introduced into the column,
flooding will result causing a loss in efficiency.
As seen in Figure 7-9, as sample size increases for
a given column size, column efficiency remains somewhat constant up to a point, after which column efficiency begins to drop rapidly.

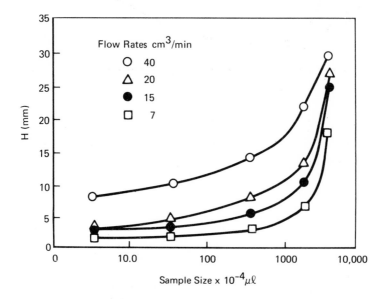

Figure 7-9.   Sample size is restricted by column size
and column type (packed or open tubular.)   As sample
size increases for a given column, efficiency remains
relatively constant up to a point at which efficiency
drops rapidly.   (Courtes   J. Gas Chromatog.) (7)

Very often, in formulating a chromatographic system,
highest efficiency and resolution are not the only
factors to be considered.   High resolution gas chromatography can be slow and requires smaller sample sizes.
These factors may not be compatible with the analyst's
needs.   Therefore, the chromatographer may sacrifice
some resolution by increasing the flow rate in order
to speed the analysis.   He may choose to use a larger
diameter column in order to increase his sample size.
Low liquid loaded columns do provide the best resolution, but, unfortunately they can deteriorate rap-

idly, particularly if large volume samples are used. The needs and requirements of each analysis must be considered in designing the best chromatographic system.

With just these few terms and quantities, plus a few more discussed later, an understanding of the individual GC systems can be readily accomplished. These quantities will be referred to several times in later chapters, and it is recommended that they be understood fully before continuing.

References

1. L.S. Ettre and A. Zlatkis Ed., The Practice of Gas Chromatography, Wiley, N.Y., (1968).

2. A.I.M. Keulemans, "Gas Chromatography", 2nd. ed., Reinhold, New York, (1959).

3. W.R. Supina and R.S. Henley, Chemistry, 37, 12 (1964).

4. Private Communication, M.T. Jackson, Jr.

5. F. Bauman and J.M. Gill, Aerograph Res. Notes, (1966).

6. J.J. VanDeemter, F.J. Zuiderweg and A. Klinkerberg, Chem. Eng. Sci., 5, 271 (1956).

7. A. Zlatkis and J.Q. Walker, J. Gas Chromatography 1, 10 (1963).

CHAPTER 8

PNEUMATIC AND SAMPLE INTRODUCTION SYSTEMS

Routine maintenance and operating principles of the pneumatic system and the sample injection system of a gas chromatograph are discussed in this chapter. Malfunctions in the pneumatic injection system can mask problems in other systems of the chromatograph; thus, familiarity with possible problems in these systems is essential when troubleshooting a particular instrument malfunction of sympton.

The pneumatic system generally consists of the carrier gas supply, pressure regulator and/or flow controllers, the column and all connecting tubing. The sample inlet system usually consists of a carrier gas preheater and the sample flash-vaporizer. A stream-splitter following the inlet system must be used when capillary columns are employed, as the small inner diameter capillary columns, usually 0.01-0.02 inches in diameter can accept only 1/100 of the microliter sample volumes generally used for analysis with packed columns. A block diagram of the pneumatic and sample inlet systems is shown in Figure 8-1.

Following are brief descriptions of how the pneumatic and sample inlet systems function and suggestions for routine maintenance of the various components.

The carrier gas is usually supplied from a cylinder (Figure 8-1, component A) with the initial pressure of 1800-2400 psig. The cylinder is fitted with a double-stage pressure regulator that allows pressure regulation between 10-250 psig. at the 2nd stage outlet. The carrier gases generally used are helium, nitrogen, hydrogen, or argon. Carrier gases are available commercially in various grades of purity. Research-grade or "GC-grade" gases are preferred because they have been refined to contain a very low level of contaminents. Contaminents commonly found in such gases are water, N_2,hydrocarbons, CO_2, and inert gases. Although nominal concentrations of

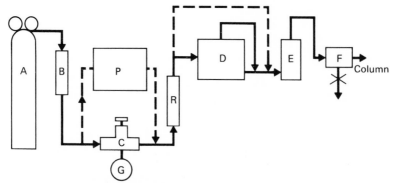

A = Carrier gas supply (usually helium or nitrogen),
    with pressure reducing regulator.

B = Carrier gas filters.

C = Pressure regulator.

D = Flow controller (with flow control needle valve).

E = Sample inlet (vaporizer).

F = Sample splitter (if capillary columns are used.

V = Split control needle valve.

R = Rotometer (optional).

P = Pressure programmer (optional).

G = Pressure gauge.

Figure 8-1   A block diagram of the pneumatic and
              sample inlet systems as far as the sep-
              arating column. (1)

impurities in the carrier gas do not appreciably affect GC retention behavior, the effect of impurities
on detector stability and response can be quite serious. The magnitude of this problem increases in direct proportion to the sensitivity of the detector.
Thus, it is of utmost importance to pass the carrier
gas through a filter or absorption trap (Figure 8-1,
component B) to remove impurities. Carrier gas filters or absorption traps usually consist of a 6-12
inch length of 3/8 inch or larger stainless steel tubing filled with Type 5A molecular sieve to remove
hydrocarbon impurities and moisture from the carrier
gas. When using a good grade of carrier gas, a molecular sieve trap should be effective for at least
one year without regeneration or replacement. How-

ever, if necessary, the sieve may be regenerated by removing from the tubing and heating to 300°C for eight hours (preferably in a vacuum oven). Charcoal filters are sometimes used for removal of light hydrocarbons an activated silica gel trap is used for moisture removal from the carrier gas.

When changing carrier gas cylinders, it is important to ensure that all fittings are free of dust and dirt particles before assembly, as these materials could enter the gas stream and cause plugging of the small orifices in the flow controllers, valves, etc. Since most gas cylinders are stored out-of-doors, be especially careful that all rust particles have been brushed away from the valve area before attaching the double-stage pressure regulator. Periodic checks for leaks should be performed using a soap solution. This should always be done when gas cylinders are changed. The major sympton of a leak in this area is the abnormally rapid use of carrier gas. Do not wait too long to change carrier gas cylinders. The outlet pressure of the tank should never be allowed to drop below 100 psig, since running the tanks dry greatly increases the probability of introducing the impurities (especially water) present in the carrier gas in the GC flow system. This could result in possible disarming of the gas filters and cause intermittant, spurious spikes to show up on the chromatographic recorder as the impurities pass through the detector.

A good rule to follow with molecular sieve traps used for removal of water and hydrocarbon impurities is to regenerate or replace the sieve in the trap about about twice a year, especially if commercial-grade gases are being used for the carrier gas. When the molecular sieve traps become saturated, impurities in the carrier gas begin to leak through into the gas-flow system and usually cause intermittant spikes to appear on the GC recorder. This problem often appears to be disturbance in the detector/electrometer system. However, if the spikes are quite·random in size and time of appearance, impurities in the carrier gas should be suspected as a possible cause. Replacement of the molecular sieve trap will eliminate the problem. Another effect of water entering the gas-flow system could be the rapid deterioration of certain types column substrates, such as polyesters, especially at elevated temperatures. Thus, it is of utmost importance to use good grades of carrier gas, and to adequately scrub impurities from the carrier gases to ensure that a pure, dry gas enters the chromatograph.

Pressure regulators (Figure 8-1; component C) are used in some commercial units and in most homemade units. Such regulators allow an accurate pressure supply to the unit, especially when more than one GC is connected to a single pressure cylinder, which could result in differential pressure changes during adjustments at other GC units. These are usually regulators of the nonbleeding, sping-loaded, diaphragm type as shown in Figure 8-2. A pressure gauge is usually an integral part of most pressure regulators.

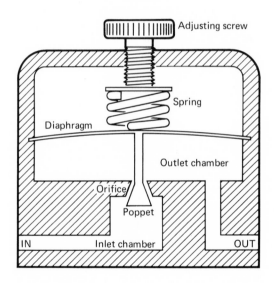

Figure 8-2   Pressure Regulator (Courtesy Veriflow Corp.).

The tube and pipe connections in and around the regulator and pressure gauge area should be checked periodically for leaks with soap solution, especially after any kind of tear down and reassembly opera-

tion. As in the case of leaks in the pressure cylinder or gas filters., leaks around the pressure
regulator and gauge will not greatly affect the GC
unit's performance, but can result in the loss of an
inordinately large amount of expensive carrier gases.
The only part of a nonbleeding pressure regulator
that can deteriorate in service and require replacement is the rubber diaphragm. A good check on the
condition of the diaphragm is to disconnect the flow-
transport line from the inlet of the rotometer
(Figure 8-1) and close it off with a plug. Adjust
the regulator until about 30 psig shows on the gauge
then back off the adjusting screw (Figure 8-2) until
no pressure can be felt on the spring-loaded diaphragm. If the gauge pressure remains constant at
about 30 psig, the rubber diaphragm and the pressure
regulator gauge system is leak-tight. However, if
the pressure on the gauge drops from the 30 psig setting, a leak is present. If the fittings in the
system all test leak tight with soap solution, a
ruptured diaphragm is indicated. The diaphragm is
easily replaced by removing the hold-down bolts along
the regulator flange, removing the old diaphragm and
replacing it with a new one. After this operation,
the above leak test should be repeated.

Rotometers (Figure 8-1, component R) are available
as inexpensive options for most commercial units,
and are used for measuring mass gas flows (calibrated in cm3/min at various inlet pressures). However,
a rotometer is only a convenience to re-establish a
given flow-rate, as flows are more accurately measured with a simple soap film meter. The main malfunction that could occur in a rotometer would be a
leak in the tube connections or the possible accumulation of dirt or excessive moisture on the inner
walls. If the carrier gas and components on the
pressure cylinder side of the rotometer are kept
clean, no maintenance should be required. However,
if some foreign material should get into a rotometer
unit, it can be disassembled, the calibrated glass
tube cleaned with a solvent and dried thoroughly.
The floats should be cleaned with soap and water and
dried thoroughly. The unit can then be reassembled,
recalibrated using a soap film meter, and put back
into service.

Differential flow controllers (with flow control
needle valve) are present in most modern GC units

(Figure 8-1, component D). A schematic of a typical
flow controller is shown in Figure 8-3. Once the
desired operating pressure has been set at the
pressure regulator, the needle valve of the flow
controller is used to set the required flow-rate.
Once the initial pressure and flow are established,
the flow through the column is kept constant during
changing pressure drops (ΔP) across the column, as
are caused by programming the column temperature
(the resistance to flow in the column increases with
increasing column temperature causing a greater ΔP

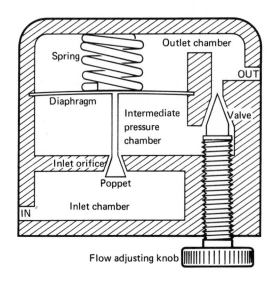

Figure 8-3   Flow Controller(Courtesy Veriflow Corp.)

across the column). The spring-loaded diaphragm
(Figure 8-3, component B) of the controller is positioned by the inlet pressure of the incoming gas
on one side and is maintained in a given position by
a balance with a combination of the spring (Figure
8-3, component C) and the pressure at the outlet.
If this balance is disrupted by a ΔP beyond the controller outlet (across the column), the diaphragm
will automatically reposition itself, causing the
controller valve to open or close, thus ensuring a
constant mass flow through the instrument. The end

effect is a constant volumetric flow of carrier gas as it exits the column at room temperature and pressure.

Pneumatic flow controllers such as the one just described have little that can go wrong except for trouble in the diaphragm area or in the comtroller needle valve. However, controllers will operate well for long periods of time if preventative maintenance is periodically performed. Most flow-controller needle valves are manufactured to close tolerances, and precautions should be taken in their care and use. Never turn the needle valve off too vigorously, this could damage the O-ring shut-off seal and the precisely machined needle. Also, when working on the carrier gas plumbing on the inlet side of the controller, take care to prevent dirt, metal chips or other foreign objects from entering the lines, as these could clog narrow apertures in the controller.

If the inlet pressure to the controller is too low, it will not compensate properly for ΔP changes beyond the outlet of the controller. At least a 10 psig is needed between the controller inlet and outlet for proper sensing of small pressure changes across the column and reliable operation. Most controllers can handle inlet pressures as high as 150-200 psig, and as a general rule, the inlet pressure should always be set as high as 60-70 psig to prevent the low-pressure differential malfunctions just described. As usual, the entire system around a controller should be periodically checked for leaks with soap solution.

The best operational check of a flow controller is to set-up the unit for a temperature-programmed analysis, and measure the total volume flow at the tail of the column or detector outlet as the column temperature increases. If the flow (cm3/min) remains constant with increasing column temperature, it may be assumed that the controller is operating efficiently. However, if flow decreases with increasing column temperature, the controller should be disassembled, all lines and orifices thoroughly cleaned, the diaphragm checked for pin-hole leaks, and any defective or apparently worn parts replaced. After reassembly, the same test should be repeated to check the performance of the overhauled controller.

It is the function of a well designed sample inlet
system (Figure 8-1, Component E) to receive the
sample, vaporize it instantaneously, and deliver the
vaporized material to the head of the analytical
column in a narrow "plug", as elution band (or peak)
widths are directly related to the injection band
width (2). Thus, effective resolution of close boiling materials could be lost by an inferior injection
system. A good example of a flash vaporizer with a
removable liner is shown in Figure 8-4 (3).

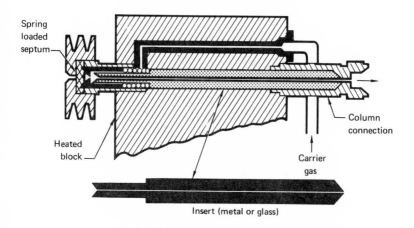

Spring
loaded
septum

Heated
block

Column
connection

Carrier
gas

Insert (metal or glass)

Figure 8-4   Typical Liquid Sample Inlet Systems (A) slider
              type (B) Rotational Type (Courtesy Perkin-
              Elmer Corp. (3)

As in all well designed vaporizers, the carrier gas
is preheated by passage through a short length of
tubing or a bored-out part of the vaporizer block
before coming in contact with the sample. The added preheater volumn also acts as a flow-surge buffer for the inlet system. To ensure "plug" flow
to the head of the column, the carrier gas enters
the inner liner at the front so that the sample will
be swept toward the head of the column, minimizing

any chance for back-diffusion or band-broadening in the vaporizer itself. This type of vaporizer is often referred to as the concentric-tube configuration. The inner liners are designed with a small internal volume (2-4 inches long by 0.028-0.030 inches i.d.) to help ensure "plug" introduction of sample at the head of the column by keeping vaporizer residence time to a minimum. However, such a high-temperature stainless-steel vaporizer can only be used with thermally stable samples that do not react with the hot metal surfaces. The same type of inlet system can be used with some unstable materials by substituting a glass liner for the stainless steel liner. Several commercial chromatographs have sample inlet system vaporizer designs similar to that shown in Figure 8-4.

The most common method for introducing liquid samples into the chromatograph is by a microsyringe. The desired sample sizes are usually between 1 to 10 microliters for analytical work, thus 5 or 10 microliter syringes are generally the most popular. The sample is introduced with the syring through a silicone-rubber septum into the vaporizer. The vaporizer is normally maintained at 250 to 350 C. At these temperatures, silicone septa begin to degrade rapidly and must be changed quite often to prevent leakage. High-temperature septa that are more resistant to thermal decomposition are now available. Also available are laminar-constructed septa, with the layer in contact with the inside of the vaporizer made from polytetrafluoroethylene (PTFE) to decrease possible absorption of components by the septum.

Malfunctions in the sample vaporizer can completely impede any reliable GC analysis. Routine or periodic maintenance of this system is a must, therefore, some of the commonly occuring malfunctions will be discussed briefly. If the inlet system is dirty, the interior walls of the system that come in contact with the vaporized sample can become covered with a thin layer of heavy coke-like material. This is especially true for samples such as gaslines that contain extremely heavy and non-volatile detergent additives that are not vaporzed at normal inlet temperatures. After a while, this material can become hardened and coke-like,

as any slightly volatile materials are slowly
swept away by the carrier gas. Such a dirty inlet
can cause several problems which are listed below.

1. Peak broadening or shifting resulting from interaction of components with active (polar) sites
on the deposit or caused by absorption effects
from the porous nature of the deposits.

2. Sometimes complete removal of polar or unstable
material will occur because of interaction with the
deposit at temperatures above 200 C.

3. Ghost peaks (peaks not related to the sample being
analyzed) will appear on the recorder. These could be
materials being desorbed from the inner lining of the
vaporizer. This problem can usually be solved by re-
tubing or replacing the inner liner of the vaporizer.

A leak in the inlet system can be one of the most
important inlet system malfunctions to the worker
who is trying to obtain consistent high-resolution
quantitative analyses, mainly because inlet system
leaks are hard to find because of the extremely high
temperature of the system. If the inlet system is
cooled to room temperature, the leak may disappear
as the metal blocks and tube fittings contract on
cooling, possibly stopping the leak. Thus, the most
expedient way to check an inlet system for leaks is
to block off the exit to the column with a tubing
plug, pressure-up the system to about 50 psig (or to
the maximum pressure at which operation will be car-
ried-out), and close off the pressure regulator
completely (as described in leak checking the pressure
regulator/gauge system previously in this chapter).
However, this technique is useful only if the pneumatic system preceeding the inlet system is leak free.
If no loss of pressure is observed the inlet system
is not leaking, and the trouble should be looked for
elsewhere in the chromatographic system. Be sure
and install a new septum becore conducting this test,
as a leaking septum would appear as a leak in any
other portion of the inlet system. Leaks can result
in the following overall symptoms: peak broadening,
loss of resolution, loss of more volatile materials
relative to heavier ones, and the appearance of ghost
peaks.

A leaky septum can produce all of the symptoms listed above. The best method of routine maintenance and operational checks that can be suggested here is to change the septum frequently. Septa are cheap compared with the cost of an individual chromatographic analysis, and a good rule to follow is to change the septum after every 10 analyses when in routine use, or weekly if used only periodically, but with the inlet system temperature maintained at the operating level. So to be safe, change the septum often. A quick and simple way to check for a leaky septum without having to cool down the inlet system is to fit a piece of 1/4 or 1/8 inch o.d. PTFE the tubing leading to a soap-film meter. The septum can then be checked for leaks by tightly pressing the open end of the tube against the septum, and checking for flow with the soap-film meter in the usual way.

Another malfunction that is often apparent in inlet system vaporizers is the problem of cold spots (i.e.- a very small area that is not heated to the same extent as the surrounding vaporizer unit). Cold spots in the inlet system are especially deleterious in the analysis of extremely high-boiling materials. The magnitude of this problem is best shown with an example.

> The GC analysis of 2,4-dinitrophenylhydra-
> zone (2,4-DNPH) derivatives of Cl-C8 alde-
> hydes was attempted using a chromatograph
> fitted with an old design inlet system
> that had about 10-inches of 1/8 inch o.d.
> stainless steel tubing following the in-
> itial vaporizer to provide an adequate
> mixing volume for the carrier gas and sam-
> ple. The complete inlet unit was heated
> to a tubing skin temperature of 350°C.
> Various columns and many sets of operat-
> ing conditions were tried to analyze a
> standard mixture of the Cl-C8 2,4-DNPH
> derivatives in benzene solution; however,
> the only derivative detected was that for
> formaldehyde (the lowest melting). On
> reappraisal of the entire system, it was
> postulated that the sample was probably
> being lost to a cold spot somewhere in
> the inlet system. Thus, to remedy the
> situation, the long, coiled inlet was re-

placed by a short glass-lined injection
port heated to 350°C. This was connected
directly to the head of the column. With
this improved inlet system, 2,4-DNPH derivatives as heavy as that for tolualde-
hyde were detected with no difficulty.
The failure of a similar old design inlet
system may have led Fedeli and Cirimele (4)
to draw the conclusion that GC separation
of 2,4-DNPH derivatives was probably
impossible.

If the samples to be analyzed are somewhat thermally
unstable, another method that should be considered
for placing a plug of sample at the head of the
analytical column is the technique of "on-column" injection, where the sample is placed by various techniques in actual contact with the column packing or
coating without the aid or dead-volume of a flash
vaporizer. Such a technique is invaluable for GC
analyses of materials that would not remain intact
when subjected to the thermal shock of the vaporizer.
In the technique usually employed, the inlet vaporizer is modified so that the head of the column can
be placed about 1/8-inch from the septum. Thus, the
syringe needle extends at least two inches into the
column for "on-column" injection. One word of
caution if this type of injection is to be used. Do
not heat the injection block above the temperature
limit of the liquid phase in the column (whether
packed or capillary). If the temperature does exceed
the limit of the stationary phase, some phase will
be stripped from the column and result in recorder
baseline drift or large, skewed peaks.

A method used for "on-column" injection into preparative size columns is through the inlet system to the
head of the column via a six-inch syringe. This
technique is generally useful for larger sample sizes
in the 20-500 microliter used in analytical preparative work.

A modified injector insert has been described by
Willis and Engelbrecht (5) for "on-column" injection
into larger diameter open tubular columns. This
technique is especially useful for analyses of
wide-boiling, temperature-sensitive mixtures. Sample
sizes used are generally in the range of 0.1 ml or
smaller.

96

A possible bad effect on resolution and general column performance when using any of the on-column injection techniques is excessive peak tailing which
may result from adsorption of the sample on uncoated
solid support or column walls. This especially becomes a problem after the stationary phase at the
head of the column caused by constant disturbance of
the packing with the syringe needle occurs.

Other types of sample inlet systems that must be
considered are those used for sampling gaseous
materials. In areas where the gases to be sampled
are under low pressures, a gas-tight syringe of the
desired size (usually 0.1 - 5.0 ml) is employed with
the usual septum-type inlet system.

If gas samples are to be taken repeatedly in areas
from which quantitive data must be derived, a gas
sampling valve, fitted with the desired size of
sampling loop, is often used. Several commercial
gas sampling valves are available, and typical configurations of these valves are shown in Figure 8-5.

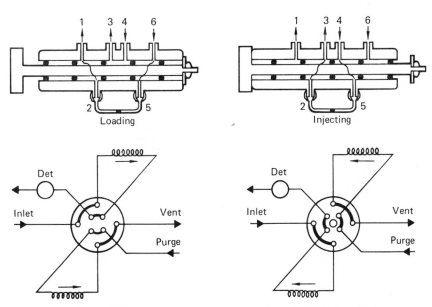

Figure 8-5   Flow Schemes of Various Sample Valves
              (Courtesy Varian Aerograph)

Some operate on the slider with O-ring principle
(Figure 8-5A), while others, such as the Carle valves,
operate by rotation of various flow paths (Figure 8-5B).
The major goal when using gas sampling valves is generally good sample size repeatability. This is
well controlled by filling the sampling loop at the
same pressure for each analysis. In addition, most
gas sampling valves may be heated to about 200°C,
eliminating any chance for condensation of traces
of heavier materials in the sampling loop or connecting tubing. This adds to repeatability.

An important item in gas sampling valve use is leak-
free operation. This is usually obtained by frequent
replacement of O-rings in the slider-type valves,
and good clean-outs at 200°C+ for the Carle-type
valves. Cleanliness of the sampling loop is another
important factor. This can be maintained by occasional
removal and solvent flushout, followed by thorough
drying of the loop before reinstallation.

Another method of gas sampling that should be considered is the technique of encapsulating a gas
sample in a small length of indium tabing (6). The
indium rapidly melts at a fairly low temperature
(157°C), and will instantaneously release an encap-
sulated gas sample. The gas in question may be
sampled by flowing it through a short length of
indium tubing. Then, using a pair of pliers, the
exit and inlet of the tube are crimped off, resulting
in a leak-tight seal, with the gas to be analyzed
securely trapped inside the indium tubing. This
technique can eliminate, in many cases, the use of
steel sample bombs which sometimes repture or leak
because of inadequate venting. The indium capsule
filled with the gas to be analyzed is then introduced
into the heated inlet of a chromatograph via a
grooved pluger assembly with a high-temperature O-ring
seal as shown in Figure 8-6. Once in the heated
inlet, the indium rapidly melts, releasing the gas
sample into the GC sample flow system. Although this
technique makes sampling a relatively easy job, it
is difficult to repeat sample sizes, and all work
should be done on a relative %w or %v basis.

Refering again to Figure 8-1, (component F), a sample
splitter is necessary if 0.01 or 0.02-inch i.d. capillary columns are used, as capillary columns will
only efficiently separate about 1/100 or less of

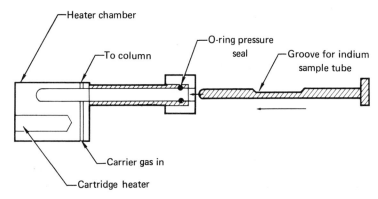

Figure 8-6   Indium Tubing Sampling System (Courtesy
              Hydrocarbon Processing)(6)

normal sample charges for packed columns, resulting
in only about 10 micrograms of total sample entering
the capillary column.   A common design for a dynamic
stream splitter is shown in Figure 8-7.

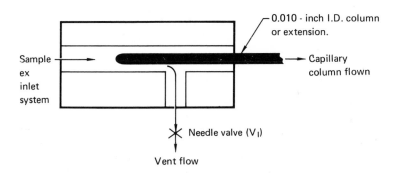

Figure 8-7   Typical Dynamic Stream Splitter.(1)

The inlet of the capillary column (or more generally, a short piece of uncoated 0.01-inch i.d. capillary tubing) is placed in the flowing gas stream ahead of the splitter vent port as shown in Figure 8-7. The ratio of sample entering the column to that being exhausted through the vent is proportional to the flow-rate ratio, or split ratio.

$$\text{Outer Split Ratio} = \frac{\text{Vent Flow}}{\text{Column Flow}}$$

The split flow is readily adjusted by a good stainless steel needle valve (V1 in Figure 8-7). To prevent any profraction or condensation of heavier materials in the sample relative to lighter components, the splitter and transfer lines must be maintained at about 300°C. Such nonselective behavior is often referred to as the "linearity" of the sample splitter (7,8,9).

A complete checkout of the stream-splitter is difficult because of the high temperature at which it must be operated. The best way to quick check a splitter is to determine its linearity for a wide-boiling standard mixture, employing different sample sizes. If, previously, a splitter had been determined linear for a given mixture, and then, under the same operating conditions behaves in a nonlinear fashion, there is a leak or deposit problem in or around the splitter and steps should be taken to clean and retube the splitter system. Since the entire system system should be made of the same type of stainless steel, a leak-tight system at room temperature should remain leak tight at 300°C as all components have the same coefficient of thermal expansion.

To protect a column that contains stationary phase material with fairly low temperature limits, the column head-pressure can be programmed to speed the elution of heavier compounds once the temperature limit of the stationary phase has been reached, (10, 11, 12). If a pressure (or flow) programmer (Figure 8-1, Component P) is in use, the differential flow controllers must be taken out of the system or they would compensate for any flow increases made by the flow programmer. The pressure programmer should be connected directly to the inlet system as shown by the dashed lines in Figure 8-1. A typical piping diagram is shown in Figure 8-8.

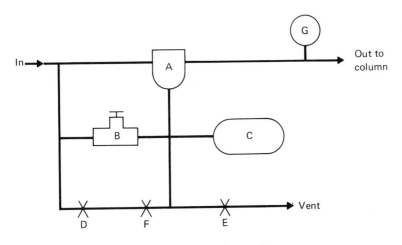

A = Pressure (flow) controller.
B = Initial pressure regulator.
C = Buffering vessel.
D = Start valve (toggle).
E = Vent valve (toggle).
F = Program-set valve (needle).
G = Column head-pressure gauge.

Figure 8-8   Pressure Programmer Flow Scheme
(Courtesy Analabs, North Haven, Conn.)
(13)

Refering to Figure 8-8, the initial column flow is
set by the pressure regulator with the start and
vent valves closed. The pressure (flow) program rate
is controlled by adjustment of the needle valve. To
begin a program, the start valve is opened. The carrier gas leaking through the needle valve builds up
a pressure in the buffer vessel that actuates a
valve in the buffer vessel that actuates a valve
in the flow controller (as discussed previously
in this chapter), and increases the column head pressure. The buffer vessel combined with the accurate
needle valve constitutes a pneumatic time constant (13).

Periodic maintenance of flow programmers is mainly
recalibration since they are entirely pneumatic devices. However, if dirty carrier gases are used,
moisture and/or particulate matter could clog small

orifices, causing erratic behavior (i.e.-deviation from previously calibrated flow programs). The same principles of cleaniness and periodic leak checks described earlier in this chapter should also be used with the flow programmers. A word of caution concerning the use of flow-programmers: determine whether the detector being used is sensitive to the flow-range being employed (14). If so, nonlinear results will be obtained unless added carrier gas is used after the column in order to make the column flow change small compared with the total gas flow through the detector.

An example, illustrating the effect of flow programming on the time required to separate a mixture is shown in Figure 8-9. The mixture consists of C_9 to C_{14} unsaturated hydrocarbons with small amounts of saturated hydrocarbons. The isothermal (125°C), constant flow (12 cm^3/min) separation of the hydrocarbons A. required 14 minutes (see Figure 8-9A). Note that the trace saturated hydrocarbons are barely seen. Figure 8-9B is a chromatogram of the same mixture (flow rate 12 cm^3/min), separated at a temperature programmed rate of 10C°/min. The analysis time is 7 minutes, and the alkanes are somewhat more pronounced in Figure 8-9B than in 8-9A. Figure 8-9C illustrates the flow programmed separation (rate of increase: 0.042 atm/sec) of the mixture held isothermally at 125°C. Again, the separation requires 7 minutes, however, the saturated peaks are even sharper than with temperature programming. Moreover, the areas of the higher boiling, i.e. C_{13}=and C_{14}=, are much larger than those shown in Figure 8-9A and 8-9B. This phenomenon is a direct result of the forementioned change in flame ionization detector sensitivity as the ratio of hydrogen to carrier gas (Helium) changes during the analysis. The change in sensitivity may or may not be an advantage to the analyist, depending upon the particulor analysis being conducted. This change in sensitivity associated with flow programming can be circumvented by using a carrier gas mixture of 60% N_2 and 40% H_2 (14).

In this chapter we have described the operating principles of the individual components that make up the pneumatic and sample inlet systems of a gas chromatograph, and discussed routine (or preventative) maintenance procedures for each component in these two systems. Since malfunctions in the pneumatic or sample inlet systems can appear as problems in other

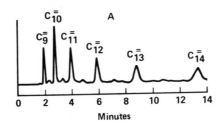

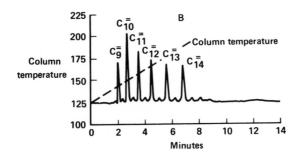

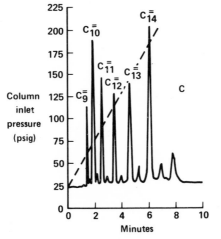

Figure 8-9   Chromatograms for the separation of a
mixture of unsaturated hydrocarbons.
(A) Column operated isothermally at 125°C
with a constant flow of 12 cm³/min,   (B)
Column temperature programmed at a rate
of 10°C/min with a constant flow rate of
12 cm³/min, and (C) Column operated isothermally at 125°C while flow programming
at a rate of 0.042 atm/sec.   (Courtesy
Amer. Lab.)

103

instrument systems, a thorough understanding of the operation of the various components in these two systems, as well as symptoms they can produce when not operating properly, is essential.

References

1. J.B. Maynard, Private Communication, 1971.

2. S. DalNorgare and R.S. Juvet, Gas-Liquid Chromatography-Theory and Practice, 3rd, Ed., Interscience Publishers, New York, 1965 (p. 166).

3. Literature on Oven Designs for Gas Chromatography, 1967, Perkin-Elmer Corporation, Norwalk, Conn.

4. Fedeli and Cirimele, J. Chromatog., 15, 475 (1964).

5. D.E. Willis and R.M. Englebrecht, J. of Gas Chromatog., August 1967.

6. J.Q. Walker, Hydrocarbon Processing, 46, No.4, 122 (1967).

7. L.S. Ettre and W. Averill, Anal. Chem., 33, 680 (1961).

8. L.S. Ettre, E.W. Cieplinski and N. Brenner, "Quantitive Aspects of Capillary Gas Chromatography", ISA Reprints, No. 78-LA/61 (1961).

9. L.S. Ettre and F.J. Kabot, Anal. Chem., 34, 1931 (1962).

10. J.D. Kelley and J.Q. Walker, J. Chromatogr. Sci., 7, 117 (1969).

11. J.D. Kelley and J.Q. Walker, Anal. Chem., 41, 1340 (1969).

12. C.J. Wolf and J.Q. Walker, Amer. Lab., 3, 10 (1971).

13. Literature on Analabs Pressure Programmer, 1969, Analabs, Inc.

14. R.L. Levy, J.Q. Walker, and C.J. Wolf, Anal. Chem., 41, 1919 (1969).

Chapter 9

Pyrolytic Inlet Systems

Introduction

Pyrolysis is not yet used by the majority of chromatographers. The technique of pyrolysis gas chromatography (PGC) is a rapidly growing field of chemistry and its
potential for use in the analysis of solids, polymers,
and many organics warrants its discussion here.
Primarily, a lack of interlaboratory reproducibility
has been responsible for the lack of widespread use
of PGC, but advances have been made and are being made
in this area. Although we will not treat the subject in
detail here, it is hoped that the discussion of this
topic will aid the chromatographer engaged in PGC
to maintain and troubleshoot his system.

It was not long after the introduction of gas chromatography that Davison, Slaney, and Wragg (1) combined GC with Pyrolysis to form pyrolysis gas chromatography. Pyrolysis itself had been used as long ago
as 1860 by Williams (2) to determine the structural
unit of natural rubber. Essentially, pyrolysis is
the process of heating a substance to high temperatures,
thereby causing fracturing of the substance into lower
molecular weight compounds. (In a very few instances,
pyrolysis will result in larger molecules being formed,
but these are of less importance.)

Introducing the pyrolysates (products of the pyrolysis
into the chromatograph results in a pyrogram. This
pyrogram can provide means of identifying an unknown
substance since the peaks are uniquely characteristic of the starting material (only stereoisomers
will have identical pyrograms). Also, the pyrogram
can provide information concerning the structure of
the starting material, be it a pure compound, a
mixture of compounds, or a polymer. PGC has found
use in many fields such as criminology[3], biology[4],
medicine[5], polymers[6], petroleum[7], and many other areas
of interest to chemistry.

Requirements

To obtain this information from the pyroylsis, certain requirements must be met. As with any analytical technique, the results must first be repeatable. This, of course, means that all factors determining which products are formed must be held constant. Probably the most important factor affecting the distribution of the products formed is pyrolysis temperature. Figure 1 shows 3 pyrograms of polystyrene. In the first pyrogram, the pyrolysis temperature was 425°C., in the second pyrogram the pyrolysis was 825°C., and in the third it was 1025°C. The amount of product is drastically different. Because large samples are difficult to heat rapidly and uniformly, small thin film samples are preferred. Another factor affecting the product distribution is illustrated in Figure 9-2. This figure compares sample size to the amount of products produced from the pyrolysis of methyl octanoate (8). Longer periods of pyrolysis will result in greater portions of the starting material being pyrolyzed. At first glance, this might seem to be an advantage, but seldom can any additional information be gained past the first instant of pyrolysis, and the additional pyrolysis time may only serve to cloud the interpretation of the chromatogram.

Secondary reactions are important when PGC is used in structure elucidation. When determining the structure of the starting material (i.e., a polymer), the most information can be determined from the initial pyrolysis. Secondary reactions are the result of reactions between the pyrolysates themselves, or may be caused by reaction of the pyrolystates with the interior walls of the pyrolysis chamber. While the reaction vessel may not necessarily react directly with the pyrolysates, it may act as a catalyst. Large dead volumes in the pyrolyzer promote secondary reactions. Steps should be taken to limit secondary reactions. Reaction vessels are made of non-catalytic materials such as quartz, glass and gold. In cases where more reactive materials must be used (such as platinum, iron, cobalt, and nickel) precautions should be taken to insure that these materials are not entering into reactions with the pyrolysates. The sample should also reach the chosen pyrolysis temperature as rapidly as possible. Rapid temperature rise times (TRT) may eliminate pyrolysis below the chosen pyrolysis temperature.

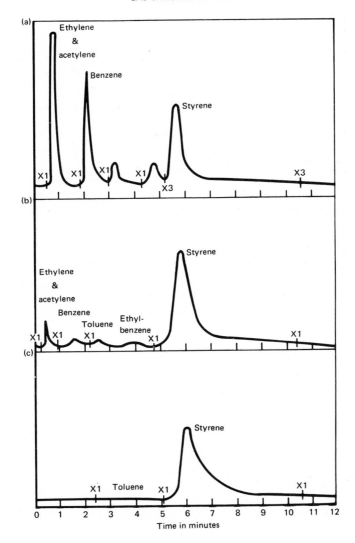

Figure 9-1 Pyrograms of polystrene at various pytolysis temperatures: (A) 425°C, (B) 825°C, (C) 1025°C column: Apiezon L; column temperature; 140°C; flow rate; 60 ml./min. Attenuation scale indicated by number in Figure. (10) (Courtesy of Anal. Chem.)

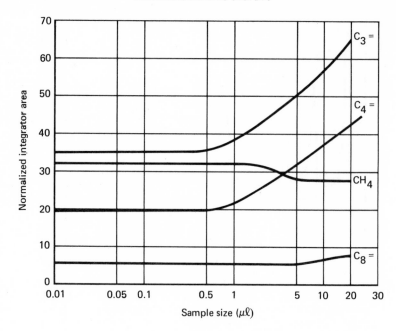

Figure 9-2   Variation in the relative yields of CH_4, C_3H_6, C_4H_8, and C_8H_{16} as a function of sample size from pyrolyzed decanoate. (8) (Courtesy of Anal. Chem.)

Summarizing , a good pyrolysis unit should provide easily repeatable procedures. It should also exhibit a fast TRT (less than one second); and, wherever possible, non-reactive and non-catalytic surfaces should be utilized in any area with which the pyrolysate might come into contact. The sample should be applied in the form of a thin film or, where possible, in the form of a vapor.

Many of these requirements have been met by today's pyrolysis apparatus. Within a given pyrolysis system, a high degree of repeatability can be established. Unfortunately, due primarily to the number of different pyrolyzers, reproducibility between different laboratories has been difficult to achieve. Essentially what is needed is a set of standard pyrolysis conditions, but little progress has been made to this end so far. Nonetheless, PGC has gained wide acceptance as an analytical instrument in several fields.

108

Instrumentation

Early apparatus in PGC consisted of   heated reaction
vessel from which the pyrolysates were collected by
means of a syringe and injected into the chromatograph.
This method was sometimes augmented by liquid nitrogen
cooling of pyrolysates to insure collection of low
molecular weight compounds(9)(10). The method is seldom
used today, having been replaced by much more reliable
and informative methods.

Current methods utilize direct introduction of the pyrolysates onto the GC column. The carrier gas flows
continuously through the pyrolyzer unit and into the
column. Pyrolyzers are classified as, either continuous mode (if the sample is introduced into a pyrolyzer pre-heated tothe pyrolysis temperature) or
pulse mode (if the sample is introduced into a cold
pyrolyzer which is then brought rapidly to the pyrolyzer temperature). Each type of pyrolyzer has characteristic advantages and disadvantages which determine its use.

Continuous mode pyrolyzers are the simplest in construction. The first of this type to be used was the
boat pyrolyzer. The sample is placed in a sample
"boat" which is resting a cool zone within the carrier
gas stream. For pyrolysis, the boat is moved mechanically (some units use a magnet) into the pre-heated
pyrolysis zone and the pyrolysates are swept into the
column immediately. Boat type pyrolyzers are rarely
used today, primarily because they are susceptible to
secondary reactions caused by large dead volumes and
a slow heat-up time.

Closely related to the boat pyrolyzer, the vapor
phase tubular reactor is still used frequently today.
As diagrammed in Figure 9-3, pre-heated carrier gas
flows into a heated injection port. Sample is injected through the injection port into the vaporizer
unit where it is vaporized, but not pyrolyzed. The
vapors then enter the pyrolyzer section which has been
maintained at the pyrolysis temperature. The pyrolyzer often consists of a length of gold tube which
limits catalytic reactions and prevents residual sample
accumulating which might interfere with subsequent
analysis. The pyrolysates are swept immediately into
the GC system travelling through tubing which is heated
to prevent condensation.

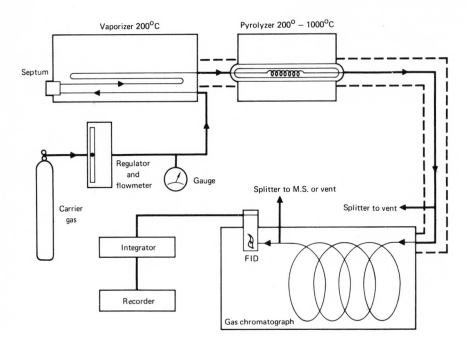

Figure 9-3   Vapor-phase Pyrolysis/GC apparatus   (17)
(Courtesy of Anal. Chem.)

Vapor phase tubular reactors afford many advantages over previous methods. No secondary reactions can occur, and the results are very repeatable. These units are very simple in design, and the pyrolysis temperature can be set at any position over a wide range of temperatures. Another advantage, especially when determining structure, is that the pyrolysates can be fairly accurately predicted on the basis of the Kossiakoff and Rice theories of free radical degradation (11). These pyrolysis units are also relatively inexpensive when compared to other types used today.

Unfortunately, vapor phase pyrolyzers are limited seriously in that they can only be used with readily volatile substances. They cannot be used, therefore, with solids and low vapor pressure liquids above C_{20}. Since the bulk of PGC work done today is in this area, other types of pyrolyzers have been developed to fit those needs.

The first of the pulse mode pyrolyzers to be used was
the resistance heated filament pyrolyzer (12-15).
The filament is enclosed in a glass tube through which
carrier gas is flowing. Preferably, the sample is
applied by dipping the filament in a solution of the
sample, then evaporating the solvent. This method
provides a thin uniform film of sample. Alternatively, a solid sample can be positioned on the
filament, but it is difficult to apply small samples
in this manner. With the sample in place, current
is applied to the wire causing it to heat up and the
sample is pyrolyzed.

These fairly inexpensive and widely used units can
be heated over a wide temperature range. Resistance
heated filament pyrolvzers exhibit some drawbacks.
Slow and sometimes difficult to reproduce temperature rise times can result in poor reproducibility.
The slow TRT also contributes to products characteristic of lower pyrolysis temperatures, making the
pyrograms difficult to interpret. Recently however,
much faster TRT's have been achieved by capacitive
boosted filaments(16,17)(see Figure 9-4). The second-

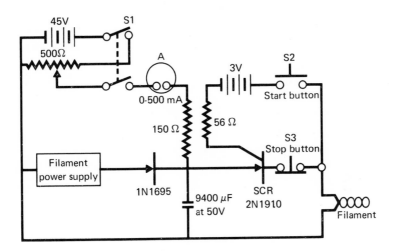

Figure 9-4   Schematic of the Capacitor Discharge
              Circuit for Rapid Excitation of Filament
              Pyrolyzers (18) (Courtesy of Anal. Chem.)

111

ary reactions with these units are nearly nonexistent and repeatability is enhanced accordingly. These units are not commercially available which prohibits their wide acceptance presently, but they will probably become of major importance in the future.

A second type of pulse mode pyrolyzer is the Curie Point Pyrolyzer. Developed in the early 1960's by Simon and his coworkers[18], these pyrolyzers utilize an interesting property of ferromagnetic materials. When exposed to a radio requency (RF) field, a ferromagnetic wire will heat rapidly. The wire attains a specific temperature (called the Curie Point) depending on the wire alloy composition and maintains that temperature until the RF field is turned off. Wires are available which will stabilize at a number of temperatures between 300 and 1000°C. Some Curie Point wires and alloy composition are shown in Table 1. A typical Curie Point apparatus is shown in Figure 9-5.

Several advantages are evident for Curie Point pyrolyzers. They can be used with solid materials, soluble solids, liquids which can adhere to the ferromagnetic wire, and some crystals and powders. The pyrolysis temperature is determined precisely and reproducibly by the Curie Point of the wire. Temperature rise times of these pyrolyzers are fast, though it depends somewhat on the RF generating unit being used. These factors contribute the high degree of repeatability which can be obtained with Curie Point pyrolysis. The simplicity of their operation indicates that they may provide some answer to the problem of repeatability between laboratories.

Curie Point pyrolyzers are not without drawbacks. Some solid materials are difficult to apply to the wires. While secondary reactions are few, they can occur in some instances. Pyrolysis temperatures are not continuously variable as they are with other types of pyrolyzers, being limited to the Curie Points of the available wires.

Several other types of pyrolyzers have been reported. Most of these are for limited use and will not be discussed here. Many, however, are included in Table 2, which compares the characteristics of the various pyrolyzers.

112

Table 1 Comparison of Curie Point Wires (6)

Manufact-urer's Specifi-cation °C	TGA C. ***	Curie Point Range °C	Manufacturer's Per Cent Comp-osition Fe-Ni-Co	X-Ray Fluor-escence Fe-Ni-Co
358*	352	±2	0-100-0	0-100-0
480*	474	±2	52-48 -0	53.5-46.5-0
510*	482	±2	49-51 -0	50.6-49.4-0
600**	597	±12	42-42 -16	42-42-16
610*	601	±4	30-70-0	29.2-70.8-0
700**	707	±12	33-33-33	33-33-33
770*	782	±4	100-0-0	100-0-0
980*	985	±5	0-60-40	39-1-60

*Philips Electronic Instruments, Mt. Bernon, N.Y.
**Fischer Labs (Varian-Aerograph, Walnut Creek, Calif.)
***Thermal Gravimetric Analyses
    (Magnetic transition point, corrected)
    Perkin-Elmer TGS-1
****Avg. of three X-Ray Fluorescence analyses
    from Siemens Crystalloflex IV X-Ray Generator
    with Vacuum Tunnel Spectrometer

(Courtesy of Analytical Chemistry)

Troubleshooting

The chromatographic system used in PGC is essentially the same as that used in conventional GC and requires no comment. Troubleshooting of the PGC system requires attention to special problems which may arise from the pyrolysis unit. The tables which follow are divided into three parts. Part I is concerned with problems involved with pulse mode pyrolyzers (Curie Point, Electrical Discharge, Cap-acitive Boosted Discharge, Laser, etc.), part II is concerned with problems which might arise from either type.

Two points should be kept in mind when maintaining pyrolysis systems. First, all sample transfer lines should be kept heated at all points. This prevents

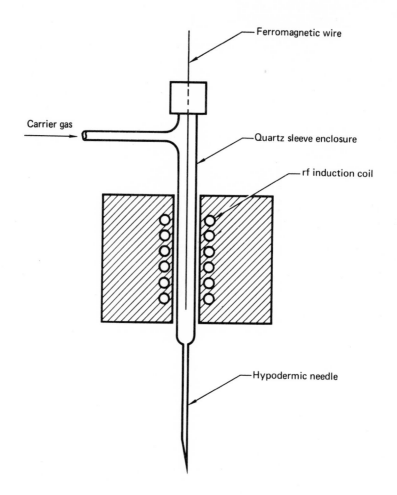

Figure 9-5   Cross-section Diagram of a Curie Point
              Pyrolyzer (18) (Courtesy of Anal. Chem.)

condensation of pyrolysates before they can enter the
column. Another important point is the frequent
occurence of leaks. Since most of the system is
maintained at high temperatures for prolonged periods,
oxidation and deterioration of seals is likely. At
extremely high temperatures it may be necessary to
replace septums every four hours. Metal tubing fittings
can develop leaks also at these high temperatures.

Chapter 9

References

1. W.H.T. Davison, S. Slaney and A.L. Wragg, _Chem. Ind._, (London) 1356 (1954).

2. C.G. Williams, _Phil. Trans._, 150, 241 (1860).

3. P.L. Kirk, _J. Gas Chromatog._, 5, 11 (1967).

4. P.M. Adhikary and R.A. Harkness, _Anal.Chem_, 41, 74 (1971).

5. E. Reiner and G.P. Kubica, _Am Rev Respirat Diseases_, 99, 42 (1969).

6. M.T. Jackson Jr. and J.Q. Walker, _Anal Chem_, 43, 74 (1971).

7. S.G. Perry in "_Advances in Chromatography_", Vol. 7, J.C. Giddings and R.A. Keller, eds., pp. 221-240, Marcel Dekker, New York, 1968.

8. W.D. Dencker and C.J. Wolf, _J. Chrom Sci_, 8, 534, (1970).

9. H. Feurberg and H.Z. Weigel, _Anal Chem_, 199 (2), 121, (1963).

10. J. Strassburger, G.M. Brauer, M. Tyron and A.F. Forziatti, _Anal Chem_, 32, 454, (1960).

11. A. Kossiakoff and F.O. Rice, _J. Am Chem Soc_, 65, 590 (1943).

12. K. Ettre and P.F. Varidi, _Anal Chem_, 34, 1543, (1962).

13. S. Tsuge, T. Okumoto, and T. Takeuchi, _Kogyo Kagaku Zasshi_, 71, 1634 (1968).

14. R.L. Levy, _J. Chromatog_, 34, 249 (1968).

15. R.L. Levy, _Chromatog. Rev._; 8, 49, (1966).

16. R.L. Levy, _J. Gas Chromatog_, 5, 107 (1967).

17. R.L. Levy, D.L. Fanter, and C.J. Wolf, _Anal._
    Chem. 44, 38 (1972).

18. W. Simon and H. Giacoggo, _Chem. Ing. Tech_, 37,
    709, (1965).

19. D.L. Fanter, R.L. Levy and C.J. Wolf, _ibid._, 44
    43 (1972).

20. R.L. Levy and D.L. Fanter, _Anal Chem_, 41, 1465,
    (1969).

Part I

Problems Encountered with Pulse Mode Pyrolyzers

Symptom of Trouble	Possible Cause	Remedy and/or Check
Poor reproducibility of relative peak areas from programs.	1. Variation in Final Temperature.	Weld Microthermocouple to filament and measure oscilloscope(20) and measure temperature.
		Measure Magnetic transition point via TGA of Curie Point wires(20).
	2. Variation in temperature rise time.	(Same as A & B above).
	3. Variation in carrier gas pressure and velocity.	Attach an accurate pressure gauge and a reliable flow controller.
	4. Variation in coil or geometry.	Figure "8" better (15).
		Curie wire must be at right angles to the radio frequency induction coil.
	5. Variation in sampling techniques.	Sample from suitable solvent 50 gm/liter(17).
		Sprinkle sample in powder state onto wire at $100°C$.
	6. Variation in sample thickness.	200-300 Angstrom units is considered ideal(17).
	7. Variation in duration of pyrolysis temperature	Short durations of 1-3 sec. ideal (6). (Measure as in A.)

8. Variations in   Pyrolysis product
   pyrolyzer   condensation can
   assembly   be reduced by
   temperature   mainaining pyr-
                    olyzer temp. from
                    100-200°C.

9. Variations in   Large dead volumes
   pyrolyzer   promote secondary
   dead volumes.   reactions.

Part II

Problems Encountered with Continuous Mode Pyrolyzers

Symptom of Trouble	Possible Cause	Remedy and/or Check
Poor reproducibility of relative peak area from pyrograms.	1. Variation in constant temperature.	Potentiometer temp. check over a period of time. If a good temp. controller is not available for the reaction, use a constant voltage transformer.
	2. Variation in temperature profile of reaction.	Plot temperature profile of reactor. Try to keep sample in a constant temperature profile.
	3. Sample size varying.	Check per cent decomposition.
	4. Variation in carrier gas flow rate pressures.	Attach a good pressure gauge and flow controller. Measure and record pressure and flow rate.
	5. Catalytic reaction of sample with reactor wall.	Check inside wall of reactor for evidence of reaction. Use gold or gold plated surface if possible.

Part III

Problems Encountered with Both
Pulse and Continuous Mode Pyrolyzers

Symptom of Trouble	Possible Cause	Remedy and/or Check
Low pyrolysis product recoveries and/or baseline drift	Leak in carrier gas system in pyrolyzer, column and/or detector	Leakcheck system at regular intervals
Pyrolysis product distributions more complex than anticipated	Catalytic Rx occuring with pyrolyzer and/or column system	Clean pyrolyzer and transfer lines and change reactive surfaces by coating with teflon or gold plating
Peak broadening	1. Large dead volumes 2. Transfer lines-pyrolyzer to GC and/or GC to detector too cool	A. Reduce dead volumes B. Add thermocouplers and/or heaters to check and maintain temperatures
High molecular weight products do not elute from pyrolyzer	Same as peak broadening causes	Same as peak broadening causes

COMPARISON OF PERFORMANCE OF VARIOUS PYROLYZERS

Characteristic Method	Sample Size	Ease of Applying Liquid & Solid Samples	Temperature Range	Pyrolysis Element Temperature Control	Catalytic and/or Secondary Reactions with Pyrolyzer
Conventional Filament and Ribbon (12)	10–1000 µg	1. Soluble in a solvent	Up to 1200°C	Continously variable	Yes
Curie Point High Power (1500 Watts) (18)	10–50 µg	1. Soluble in a solvent 2. Some Crystals and powders	Up to 985°C	Limited to 352,400,482, 597, 601,500, 707,782,985°C	Yes
Curie Point Low Power (30–100 Watts) (6)	10–50 µg	1. Soluble in a solvent 2. Some crystals and powders (Heat wire 100°C)	Up to 985°C	Limited to 352,400,482, 597, 601,500, 707,782,985°C	Yes
Capacitive Boosted Filament (17)	5–10 µg	1. Soluble in a solvent 2. Heat wire to 100°C	Up to 900°C	Continuously variable	No

121

Method	Sample Size	Ease of Applying Sample	Temperature Range	Pyro Element Control	Catalytic Effect
Tube Reactor and Boat (15)	1 μg to 5000	All forms Liquids Solids	Up to 1500°C	Continuously variable	Slight
Vapor Phase (8)	.001 to 10 μg	Volatile materials only	Up to 800°C	Continuously variable	No
Laser (19)	500 μg or Larger	Dark sample or media Such as Carbon	Estimated 1200 to 1500°K	Not controlled	Yes with Graphite
Electric Discharge (15)	200 μg	All Forms		N/A	Slight

Characteristic Method	Temperature Rise Time	Available on the Market	Repeatability	Disadvantages	Advantages
Conventional Filament and Ribbon (12)	2-10 sec	Yes	Fair	1. Slow Rise Time 2. Aging of metal 3. Solid material hard to apply	1. Most Widely used 2. Wide temp. range 3. Purchased or built 4. Simple to apply sample
Curie Point High Power (1000 Watts) (18)	40-130 Milli-seconds	Yes	Good	1. Solid material hard to apply 2. Pyrolysis chamber must be heated	1. Fast rise time. 2. Accurate temp 3. Good repeatability 4. Can be purchased.
Curie Point Low Power (30-100 Watts) (6)	0.5 -2 sec.	Yes	Fair	1. Solid material hard to apply 2. Pyrolysis chamber must be heated 3. Slow rise time	1. Accurate temp 2. Fair repeatability 3. Can be purchased
Capacitive Boosted Filament (17)	15 Milli-seconds	No	Good	1. Solid material must be soluble	1. Fast rise time

Method	TRT	Available	Repeatability	Disadvantages	Advantages
Capacitive Boosted Filament (cont.)				2. Must use small sample sizes 3. Cannot be purchased	2. Wide temp. range 3. Good repeatability
Tube Reactor and Boat	30–50 sec	Yes	Fair	1. Slow rise time 2. Requires large samples 3. Does not lend itself to capillary columns	1. Any material liquid or solid 2. Widest temp range 3. Wide sample size range 4. Can be purchased 5. Results can be compared to kinetic studies
Vapor Phase	N/A	Yes	Good	1. Only volatile materials	1. Good repeatability 2. Wide temp. range 3. Can be correlated with theory
Laser	Estimated 10 sec	No	Good (Blue glass)	1. Must use colored agent to pyrolyze 2. Complicated	1. Pyrograms may be simple 2. Superior rise time

Method	TRT	Available	Repeatibility	Disadvantages	Advantages
Laser (cont.)				3. Large sample size 4. Equilibrium temp high than other pyrolyzers 5. Laboratory hazards 6. Cannot be purchased 7. Complex fragmentation process	3. Not necessarily thermal degradation.
Electric Discharge	50-10 sec	Yes	Poor	1. Slow rise time 2. Large sample required 3. Poor repeatability	1. Any material liquid or solid 2. Wide temp range 3. Slow rise time

Chapter 10

Gas Chromatographic Column Ovens
and Temperature Controllers

The purpose of this chapter is to present sufficient
information to allow a worker in gas chromatography
to recognize and correct malfunctions in with the
analytical column, the column thermostat (or oven),
the column over temperature controller or programmer,
and the sample transfer lines running from the inlet
to the detector. The column system is the heart of
the chromatograph. If it is not functioning properly
the finest inlet and detector systems would be of
little practical use. The desired function of each
component will be described along with routine
maintenance procedures and actual laboratory maintenance problems.

The nucleus of any GC system lies in the analytical
column where the separation of the components in a
mixture actually occurs. The right column in GC is
one that has the proper size and stationary phase.
This column must be operated under the optimum conditions that will separate the components desired. To
select the column liquid stationary phase, "like
dissolves like" is a good general rule to follow.
Thus, nonpolar liquid phases are best for separating
nonpolar mixtures, such as paraffinic hydrocarbons,
while polar stationary phases generally provide optimum
separation of polar compounds, such as alcohols or
phenols. Table 10-1 shows some examples of stationary (liquid) phases, their molecular structure, approx-
imate useful temperature range, and the classes of
materials they would separate. (1)

The column dimensions should also be optimized for
capacity (sample size) and speed of analysis. Table
10-2 gives the features and characteristics of typical
analytical and preparative-scale columns. (2) The
capillary, 01158 cm and 0.317 cm outer diameter
columns are used for greatest efficiency. These
columns have small inner diameters, use small

Table 10-1

Stationary Phases For Analysis
Of Sample Types Shown

Liquid Films(in order of increasing polarity)

For saturated hydrocarbons and other non-polar compounds:

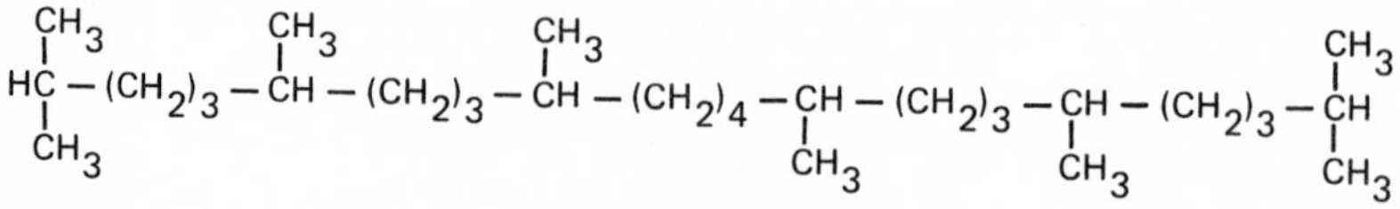

0 to 100°C

Squalane

For unsaturated hydrocarbons and other semi-polar compounds:

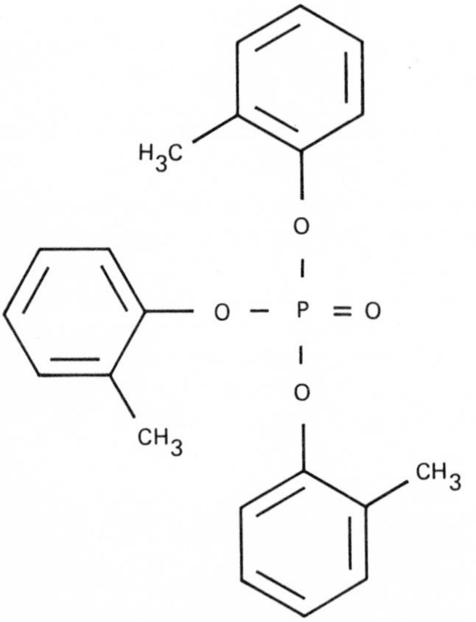

Tri cresylphosphate   20 to 125°C

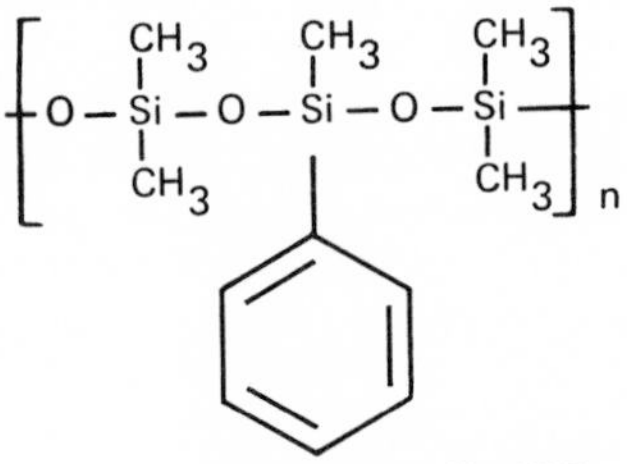

SE - 52   50 to 300°C

Dow - 710 }
Dow - 550 } −15 to 250°C

128

For alcohols, esters, ketones and acetates:

$$OH \left[CH_2 - CH_2 - O \right]_n H$$

Carbowax
20M   60 to 220°C
400   10 to 100°C

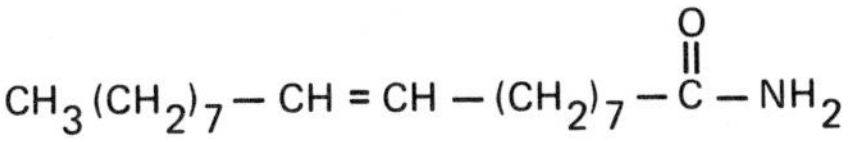

$$CH_3 (CH_2)_7 - CH = CH - (CH_2)_7 - \overset{\overset{\displaystyle O}{\|}}{C} - NH_2$$

Hallcomid M 180L   −8 to 150°C

For fatty acid methyl esters:

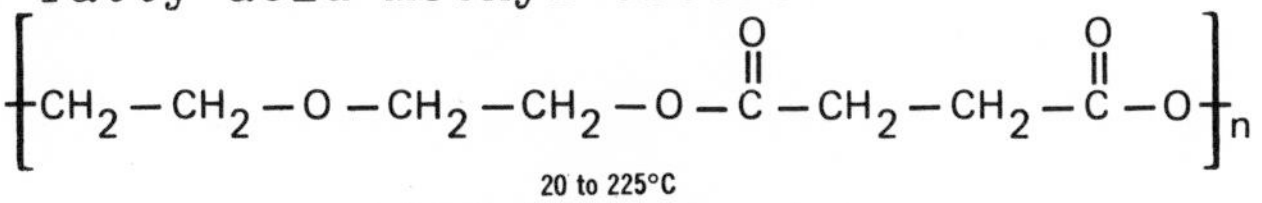

$$\left[CH_2 - CH_2 - O - CH_2 - CH_2 - O - \overset{\overset{\displaystyle O}{\|}}{C} - CH_2 - CH_2 - \overset{\overset{\displaystyle O}{\|}}{C} - O \right]_n$$

20 to 225°C
Diethyleneglycol succinate

Highly polar

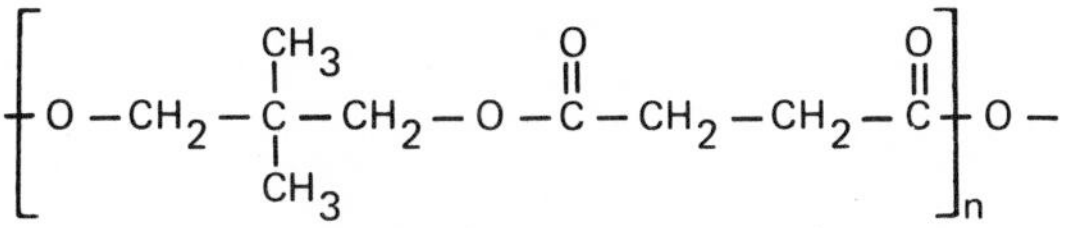

$$\left[O - CH_2 - \overset{\overset{\displaystyle CH_3}{|}}{\underset{\underset{\displaystyle CH_3}{|}}{C}} - CH_2 - O - \overset{\overset{\displaystyle O}{\|}}{C} - CH_2 - CH_2 - \overset{\overset{\displaystyle O}{\|}}{C} \right]_n O -$$

50 to 225°C
Neopentyl glycol succinate

Slightly polar

For nitrogen compounds:

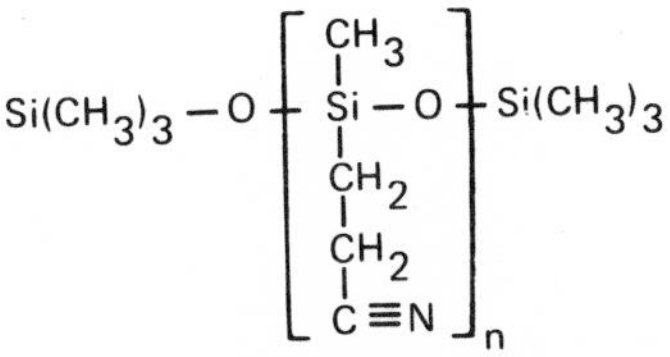

$$Si(CH_3)_3 - O \left[\overset{\overset{\displaystyle CH_3}{|}}{\underset{\underset{\underset{\underset{\displaystyle C \equiv N}{|}}{\underset{\displaystyle CH_2}{|}}}{\underset{\displaystyle CH_2}{|}}}{Si}} - O \right]_n Si(CH_3)_3$$

XF - 1150   XF - 1112
XE - 60   XF - 1125
0 to 240°C

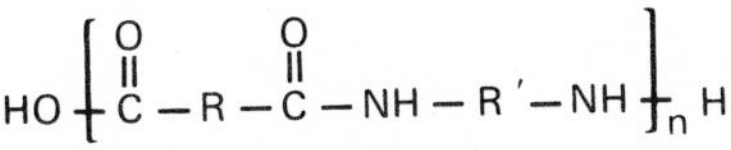

$$HO \left[\overset{\overset{\displaystyle O}{\|}}{C} - R - \overset{\overset{\displaystyle O}{\|}}{C} - NH - R' - NH \right]_n H$$

Versamid

190 to 275°C

For halogen compounds(incl. pestacides), freons and alkaloids:

$$Si(CH_3)_3 - \left[O - \underset{\underset{CH_3}{|}}{\overset{\overset{CF_3}{|} \overset{CH_2}{|} \overset{CH_2}{|}}{Si}} - \right]_x \left[O - \underset{\underset{CH_3}{|}}{\overset{\overset{CH_3}{|}}{Si}} - \right]_y O - Si(CH_3)_3$$

0 to 250°C

QF-1 (FS - 1265)

$$CF_3 \left[CF_2 \right]_n CG_3 \qquad\qquad Cl \left[CF_2\, CFCl \right]_n Cl$$

Kel F oil No. 10, No. 3   50 to 150°C   Kel F grease

Solid Support

For glycols and gases(CO_2, H_2S, CS_2, N_2O, and NO):

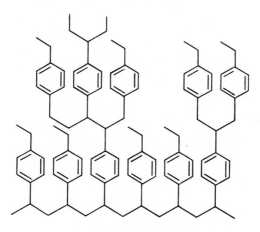

Porapak   −20 to 225°C

Table 10-2
Comparison of Column Sizes

Parameter ↓　　Size →	Capillary	0.158 cm 1/16 in.	0.317 cm 1/8 in.	0.635 cm 1/4 in.	0.952 cm 3/8 in.
Inside diameter, cm	.02-.05	0.119	0.165	0.393	0.800
Maximum Plates/m	3300	330	260	160	100
Max. Practical length, m	92	18	18	18	31
Amt. liquid phase, wt %	—	3	5	10	20
Liquid film thickness (μ)	1	5	5	10	20
Particle diameter range(μm)	—	100/120	80/100	60/80	20/40
Permeability, x 10^{-7} cm^2	200-800	1.5	2	3.5	4.0
Avg. linear velocity, cm/sec	25	10	10	7	7
Max. Sample size, μl	0.010	1.0	2.0	20	1000

particle-size solid supports, or thin liquid phase films and require very small sample sizes (1/100 ug). The 0.635 cm outer diameter columns sacrifice some efficiency for capacity in semi-preparative GC work. The 0.952 cm outer diameter columns are used when capacity, not difficult separations, is required.

Meticulous attention to detail in preparing columns can ensure good operation. Many chromatographers do not take enough time or care in column preparation, and as a result, spend considerable amounts of time and money trying to achieve difficult separations using a mediocre column. Contrary to conjecture, column preparation is indeed a science. Duplicate columns prepared from the same set of ingredients should indeed be very similar in properties. This is certainly the case, for example, in most "matched columns" prepared for dual-channel programmed-temperature GC units.

The column packing material consists of a stationary liquid uniformly coated on a narrowly sieved solid support (usually a firebrick or diatomaceous earth). The material is then loosely packed into the column employing a variety of techniques such as vibration, vacuum, etc., so that no channeling or areas void of packing occur in the column. The ideal capillary, or open, (2) wall-coated column would be one in which the entire inner column wall is coated with a uniform thickness of the stationary phase. For more details on Column Preparation see Appendix F.

With little effort, GC columns can be maintained so that they will deliver reliable analyses. During use, the average column builds up contaminants from heavy samples that, after a time, can cause deteriorating column performance and possible shost peaks on the recorder. This can often be corrected by "steam cleaning". (3) Steam cleaning can be accomplished by heating all of the GC sample system above 100°C (150°C usually preferred by authors) and injecting several 10-20 μl samples of water. If a flame ionization detector (FID is used, water should be injected through the inlet until little or no organic material is desorbed from the column by the polar steam. This results in essentially no detectable weekly FID response as the water passes through. Periodic weekly steam cleaning of a system will

132

greatly reduce the need for extensive replacement of inlet systems and connecting lines and will usually increase useable column life.

Be sure and know the temperature limitations of the stationary phase being used (see Table 10-1) and keep all operations at least 10-15°C below this critical temperature. Observance of this criterion will certainly add to useful column life and prevent the detector and other parts of the unit from contamination with the "bleeding" stationary phase. A sure sign of excessive column temperatures is not always gradual column bleed as evidenced by a slowly upward drifting baseline, but by large, broad and irregular shaped peaks. Typical large, irregular peaks are shown in Figure 10-1.

Note: Taken from actual analysis of gasoline on a 200-foot squalane capillary when temperature "mistakenly" reached 160°C.

Figure 10-1 Typical Columb "Blurps" (6)

To keep the column free of foreign materials and mositure, a constant, dry, carrier-gas flow rate should be maintained. Another aid is to store the column sealed with a small positive pressure of the dry carrier gas. This can usually be accomplished by capping the ends of the column with suitable silicone rubber plugs after a constant gas flow has

been maintained to purge the column. If possible,
a column should also be given a good steam cleaming
and high-temperature purge before storage. This
will generally eliminate foreign matter in the column
that could cause degradation of the stationary phase
during storage.

A drifting baseline can be caused by a new column not
properly conditioned (i.e., still bleeding traces of
stationary phase and solvent). To isolate the problem disconnect the column from the detector. If
the drifting stops and the recorder pen returns to
baseline, column irregularities are indeed causing
the drift and can be eliminated (or improved) by
using one, or all, of the following suggestions. If
drift continues after disconnecting the column, the
problems are electrical and will be discussed in
subsequent chapters.

> 1. Recondition the column by disconnecting it
> from the detector, increasing the carrier gas
> flow and setting the temperature near the
> stationary phase limit. Continue to condition
> the column in this manner overnight. Then
> cool it to the original operating temperature
> and pressure, and reconnect to the detector.
> If baseline drift persists, proceed to suggestion
> #2.
>
> 2. The drift could be from column contamination
> or deterioration. Try steam cleaning as de-
> scribed previously in this chapter. If drift
> continues after this treatment, the column is
> probably badly deteriorated or, for some reason,
> unstable, and should be repacked, recoated or
> replaced.

Deteriorating resolution is usually caused by
a deteriorating or contaminated column, or a
leak in the system. Quickly leak check the
system as described in Chapter 8, especially if
it has been operated under widely varying programmed-temperature conditions. A good steam
cleaning should be tried next. If resolution
remains poor, the column should be replaced or
recoated as required.

Fluctuating resolution is usually caused by an unstable condition within the column, such as channel-

ing in packed columns or deterioration of the coating in capillarys. The solution to this problem is usually replacement of the column. This problem could also be caused by leaks, dirty connecting tubing, or fluctuating or cycling column temperature or pressure.

A decrease in retention time is usually caused by column deterioration and is solved by replacement or recoating of the column. However, a false low temperature reading also could cause a loss in retention time.

Ghost peaks can come from areas other than the column, such as the septum or inlet system. However, ghost peaks can also originate in the analytical column when it is contaminated with heavy or polar materials. The best overall treatment for the appearance of ghost peaks is steam cleaning of the chromatographic system. If this doesn't solve the problem, column replacement or recoating is in order.

Peak height changes usually occur because of column changes that result in peak broadening or reduced column temperature. Thus, column condition, flow, and column temperature should be checked to solve this problem. Oxidation of the stationary phase or retention of water could also result in peak broadening.

To conclude the discussion of columns, the configuration in which the column is coiled should be considered. The two most common coiling configurations for packed columns are shown in Figure 10-2. With the carrier gas continually flowing through the column, the force of gravity pulling downward on the column packing could result in channeling, with the observed effect being a noticeable loss of selectivity and resolution. Thus, the vertically coiled configuration (see Figure 10-2B) would be the best to use, as the packing could only channel at the top and bottom of the coils. The horizontally coiled column (see Figure 10-2A) could exhibit channeling throughout the length of the column. Because of the generally short length and adequate distance between coils of packed columns, uniform heating of the coils offers no difficulty. Often, however, the choice of vertical or horizontal coiling is made depending

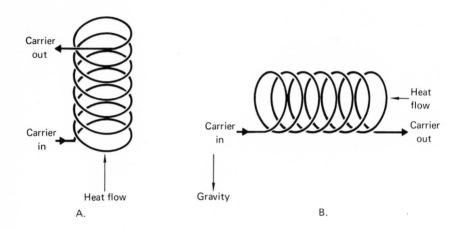

Figure 10-2   Most Common A (and Ideal B) Packed Column
              Configurations (4)

on the type of oven design the columns are to be
used with. Some column ovens require horizontal
coiling so that the heated air will move about the
column in a cycloidal fashion or will move axially
through the column coil. (7) Other ovens require
the column to be coiled vertically to obtain the same
thermal effects. A suggested minimum coil diameter
to which columns should be wound has been previously
described.

The configuration in which capillary column is coiled
will not result in inner disturbances such as channeling; however, the correct method of coiling capillaries is paramount in obtaining even heating.
Capillary columns are usually wound on some sort of
mandrel, from a common tin can to a spaced-mandrel
type of coil. Winding a 100-foot capillary on a
one piece mandrel such as a tin can is a neat and
space saving technique; however, with two to four
layers of the column coiled on the can, uniform

heating during temperature-programming is difficult
to obtain. At a given isothermal temperature setting,
the interior coils of the column will eventually
approach thermal equilibrium with the outer coils by
heat conduction and fairly even heating will result.
However, if the temperature is to be programmed, there
is little chance the inner coils could be at the same
temperature as the outer coils as the heat input rises.
To prevent such thermal gradients across the column,
each coil should be separated by an air space to
allow good circulation and obtain maximum benefit of
column temperature programming. The ideal capillary
coiling configuration is shown in Figure 10-3. Sometimes referred to as "basket-coiled", (8) such a

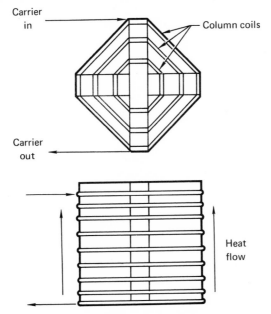

Figure 10-3   Mandrel Winding of Capillary Columns (8)

configuration allows fairly even heating of each
coil of the extremely long (up to 1000 feet in some
cases) capillary columns. The preparation of a
similar system has been described previously in detail.(9) Such spaced-mandrel winding should ideally
provide the best possible uniformity in heating,
assuming à uniform oven heat flow. Some capillary
chromatographers claim that loose coils of capillary
tubing held together with wire can be temperature-

programmed with good results. This may be true, assuming the coils are not too tightly wired together; however, there invariably exist some temperature gradients where the coils contact one another. In general, the column, whether packed or capillary, should be coiled to prevent contact of adjacent column coils and to enhance good air circulation around all external areas of the column.

A good column oven design should include adequate precautions to ensure that the oven compartment is not effected by other heaters, such as the inlet or detector heater. The column oven should be free from the influence of changing ambient temperatures and have a well-designed and adequate air flow system to maintain good temperature control of the column. In the actual design of commercial ovens, many of the criteria stated above are met (11,12 & 13). The generally accepted way to heat a column to the desired temperature is by use of an air-circulating oven, although other techniques can be used such as: immersion in a liquid oil circulating constant-temperature bath; stagnant air heating (such as closed ovens with no fan, or covering the column with a simple heating mantel); or by direct passage of electrical current through the metal column walls. Most air circulating ovens are generally of the designs shown in Figure 10-4. In Figure 10-4A, the

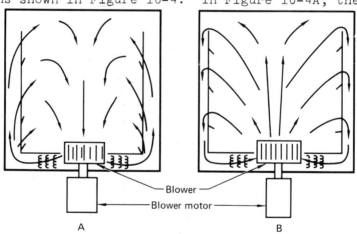

Note: Arrows inside ovens indicate air flow.

Figure 10-4   General Oven Designs (10)

138

air is blown past the heating coils then through the baffles that make up the inner wall of the oven, past the column and back to the blower to be reheated and recirculated.

The air-flow configuration in Figure 10-4B is exactly opposite that of Figure 10-4A in that the air is sucked through the wall baffles, past the heating coils, and the freshly warned air expelled by the blower. The blower drive motor should be well insulated from the oven environment, since not many electrical motors could withstand heat leaks with inside oven temperatures greater than 200°C. In general, rapid circulation of heated air within the column oven is essential. High-capacity, squirrel-cage blowers and carefully located baffles provide this, ensuring homegeneous temperature distribution, accurate thermal regulation, and rapid cooling when desired. Most column ovens are constructed of low-mass stainless steel to permit rapid heat-up and cool-down, and allow a uniform temperature to be maintained of ±1°C of the isothermal desired value and ±2°C of the desired temperature during programmed operation. (10) There is usually little influence on the oven temperature from detector or inlet system heaters above 50°C, and programmed heating is usually quite linear. Another desirable feature of better ovens is that they are in separate modules from the electronic components. This is not to protect the column oven itself, but rather to protect the electronic components from heat that may escape from the column oven during high-temperature operation.

A GC column may be heated by passage of current directly through the metal walls of the column to produce the desired column temperature (11). For this type of operation the column is usually placed in a draft-free enclosure. However, this method of column heating requires electrical insulation from other chromatograph parts. Another method of directly heating the column is by wrapping it with heating tape or resistance wire; however, this is generally not desirable, since column cool-down takes an inordinately long time.

For essentially trouble free operation, the column oven should be checked periodically, and minor maintenance performed. At least once every six months (especially under heavy usage), the blower moter (if not of the sealed bearing type), the oven lid, and other moving parts should be lubricated. To prevent sudden power failures, periodic checks of the power connection to the oven heater should be made. Also, the actual heating wires should be periodically checked for signs of corrosion or deterioration, especially if the GC unit is situated in a somewhat corrosive atmosphere. Replacement of an obviously corroded or eroded heater wire may save time since a complete burn-out could occur during an analysis.

Another factor that could contribute to poor oven operation is a failure of the temperature-sensing thermocouple of the controller. Be sure that it is firmly installed in the proper place in commercial ovens. In homemade ovens, it is paramount that the thermocouple be exposed to the oven air in an area adjacent or near the column to obtain an accurate indication of the actual column temperature. Some workers attach the sensing thermocouple to the column itself, but this is not as accurate a technique for sensing total column temperatures, unless you can be sure the spot on the column to which the thermocouple is attached is indeed representative of the whole column. In addition, if operating consistantly at high temperatures or in any type of corrosive atmosphere, the thermocouple wire should be watched closely for corrosion damage which could result in false temperature indications.

It should be beneficial to most chromatographers to consider a few oven system problems that have been encountered. For example, what is the procedure to follow when you set the column oven to heat to 150°C and nothing happens. First, check the obvious and make sure all electrical connections to the oven from the controller are intact. Make sure that no power cords have been accidentally pulled out or vibrated loose. If all of the connections are intact, and the theromcouple is reading the correct temperature (probably room temperature in this case), use a simple AC voltmeter determine power is present at the temperature controller output. Also, check the oven heater for continuity to establish that the column oven does not have an open heating element. If all of these tests are negative except

that no power is coming from the controller, repair
of the temperature controlling unit is required.
If an open element is found in the oven heater wires,
this is generally quite easily replaced in the laboratory with conventional spring-steel heating wire.
For troubleshooting of silicone controlled rectifier
circuits, see Appendix D.

Fluctuating-column-temperature can be a much more
complex problem. One of the most likely causes of
this phenomena could be an intermittant leak to
ambient air where room air could cause changes in
oven temperature. A faulty or corroded temperature
sensing thermocouple could cause temperature
fluctuations, especially if programming the temperature might cause distortion of the thermocouple
itself. Generally, a good cleaning and resoldering
of the thermocouple tip will take care of this
problem. Another possible cause of fluctuating
column temperature could be trouble in the oven
temperature controller. Probable solutions to this
difficulty will be discussed in the next section.
Cycling or fluctuation in column oven temperature
can also be caused by fluctuating oven blower
speed, caused by fluctuating line voltages.

Ovenheating has ruined many analytical columns.
To prevent this, an overheat protection system is
necessary. The most common is thermal fuses made
of alloys that melt at certain predetermined temperatures, completely shutting off power to the oven.
Most modern controllers and programmers have temperature limit shut-off switches which consist of
a microswitch attached to the pyrometer. When the
sensed oven temperature reaches this pre-set level
(selected by setting the maximum temperature set
point on the pyrometer) the microswitch opens, cutting power to the oven. Such a system, operating
from the measured column temperature is superior
to any system operating through the programmer or
controller readout temperature. Usually this upper
limiting switch has an accuracy of ± 10%.

Temperature overshoot when approaching an isothermal
point of operation is usually caused by poor air
circulation in the oven, especially if the temperature required is considerably above that at which
the column had been operating. This overshooting

effect is much more pronounced with on/off full-power type controllers than the power proportioning type. If a stagnant air oven (such as a heating mantel) is being used, temperature overshoot when changing to a higher column operating temperature can be minimized by a low power setting on the controllor, or if a programmer is being used, slowly programming to the desired temperature.

Temperature overshooting is also caused by a stalled oven blower. This is usually attributable to failure of the blower moter, caused by excessive heat, worn bearings, etc. Use a voltmeter to determine if the motor is recieving power; if not, check all connections and make necessary repairs. This is easily done from schematics supplied with most commercial units. If all these tests show no malfunctions, the moter must be replaced or repaired and except for lower temperature, isothermal operation, the chromatograph is "down for repairs". Removal of most blower motors is usually quite easy after the blower has been removed from the shaft.

Poor oven lid seals would result in temperature cycling and inability to hold high temperatures. The obvious solution to this problem is replacement of the faulty seals.

One of the most common problems with column ovens is that of not being well enough insulated from other heated components of the chromatographic system (vaporizer, detector, splitter, etc.) This is especially true if the column must be operated at 50°C or less, as these components usually must be operated at least at 200°C for good performance. Even with the presence of large thermal barriers some heat is always passed by conduction through the connections to the column. Thus, a completely isolated, yet usable oven is hard to achieve.

All of these different column heating devices should be interfaced with an accurate and sensitive temperature controller. The controller should continually sense column oven air temperature and supply the amount of power required by the oven heating coils to obtain the desired heat input. In general, three types of temperature controllers are used; the constant voltage unit, the on/off full power unit, or the power proportioning unit. The most common type of constant voltage controller is the conventional powerstat. In early GC units, the column temperature was often crudely controlled by a powerstat; however, without proper temperature sensing and feedback devices, the column temperature could not be well controlled. The earliest feedback type of unit was the full-power, on/off type of controller. A temperature setting is selected, the temperature, measured accurately by various types of sensors, the sensor would supply feedback information to the power supply for the oven heater, either calling for heat or not. Thus, the oven would be kept at a given temperature by cycling the heater on and off at full power as required by the sensor. However, with this type of control, small cycles in temperature sometimes develop, resulting in inconsistent column operation. The constant voltage type of controller supplies a constant voltage at a reduced level to provide the desired temperature (i.e., reduced from full line voltage). The unit is isolated from fluctuations in line voltage by an isolation transformer. Such controllers are usually quite accurate if adequate insulation of the column from ambient air changes in provided. Power proportioning controllers are generally more sophisticated forms of constant voltage units, employing solid state circuitry and rapid time-constant sensing circuits.

If the temperature of the column oven is to be programmed, the programmer should instantaneously supply increasing power to the heating coils on command of the controlling unit. There are also several types of power and control units for programming the column oven temperature, from simple mechanical drive units that supply heat required by cycling full oven power to advanced power proportioning models, where heating rates are generated electronically by complicated solid state circuitry. The more popular and less expensive mechanical drive programmers have rates selected by various gear ratios, the gears being driven by a constant-speed synchronous motor. After calibration procedures, so that the temperature read-out for the digital mechanical system is consistent with the measuring thermocouple or sensor, a feed-back loop allows supply of the amount of power required to establish

the programming rate. Electronic programmers depend
mainly on different solid state time-logic RC
circuits to increase the power to the column oven
to supply the desired heating rates.

Temperature controllers and programmers are relatively
easy to maintain, especially if they are of the new
solid-state variety, where operational heat, such as
found in tube-type units, is absent. In general,
a calibration check monthly or as prescribed by
the manufacturer is sufficient to determine efficiency
of operation. Calibration proceedures are well defined in operations manuals for various types of
programmers and controllers, and will nto be discussed
here. However, one quick way to check the accuracy
of the sensing thermocouple circuit is to immerse
it in boiling water and adjust the meter to 212°F
(100°C). If the unit is fitted with a subambient
unit, check the thermocouple in ice (32°F or 0°C),
and inspect the coolant relay operation. If the
unit has a mechanical drive for the temperature program (servo-motor plus a gear system), periodically
check the gears for signs of wear or missing teeth.
In a mechanical drive unit, never force-mesh the
gears together while the serve-motor is on, as this
will generally result in gear damage. A little care
when selecting programming rates in gear driven
units will go a long way toward prolonging the life
of most programmer gear systems.

Temperature controllers and programmers result in a
good portion of the problems encountered in the column oven system. Temperature fluctuations are
frequently caused by faulty or failing controllers
or programmers. If the rest of the oven unit checks
out as described previously in this section, then the
cause for this problem obviously lies in the controller. Some suggestions might be:
1. Relocate the thermocouple in the oven to
    an area of greater exposure.
2. Recalibrate the unit, as synchronization
    between the required temperature and actual
    temperature be off.
3. Thermocouple circuit may be picking up elect-
    rical "hash". This is evidenced by vibra-
    tion of the pyrometer needle. Re-route or
    shield the thermocouple leads.
4. Check reference voltages to and from the

      sensing units as described in most manuals
      (especially the output of SCR power-proport-
      ioning systems).

5. The sensing unit could be inadequately damped,
   and the problem could be cured by additional
   damping.

If the oven will not heat the trouble could lie in
the controller. First of all, check to see that
the temperature setting is high enough to bring power
on. Check the thermocouple circuit. Finally check
the output voltage of the controller as described
in the user's manual.

Overshooting the selected temperature is often caused
by poor calibration or by the unit being overly
damped; thus, not sensing the temperature changes
rapidly enough to control at the selected temperature. Solve this problem by decreasing the damping.

When programming the temperature, a common malfunction observed is stepwise jumps in temperature. This
is usually caused by one of two things: if mechanical drive, the gears are slipping, or the unit
is either not calibrated correctly or is overdamped.
These are all easy items to rectify.

A nonlinear program rate usually suggests the serve-
motor for the mechanical drive unit is malfunctioning, or that there is an intermittent leak in the
air bath thermostat. In power proportioning units,
electrical malfunctions within the programmer are
possible. This usually is indicative of faulty
voltages being sensed by the controller units, and
requires expert attention.

One last item to consider in the column oven system
are sample transfer lines between the vaporizer and
column and from the column exit to the detector.
For optimum operation of any GC system, this connecting tubing should have as small a volume as is
practicable, as all connecting lines constitute
dead volume. In most commercial and home-made units
these connecting lines are usually too large, too
long, and especially too cool. Only 1/4 inch of
unheated 1/8 inch OD stainless steel tubing can
result in a 25°C temperature drop. Transfer lines
larger than 1/8 inch OD should never be used.

Instead, well-heated, large bore capillary tubing is preferred. Good GC design demands short and efficient sample transfer lines.

Routine maintenance of sample transfer lines is equally as important as the other areas of the column system. When the system is cool, and with installation of each new column, all connections should be checked for gas leaks with soap solution. Steam cleaning of the column as described previously should also do an adequate job of removing adsorbed materials from these lines. If very heavy or corrosive samples have been run, such as pesticides or alkyl-lead compounds, remove these lines periodically and flush with solvents. If large deposits are on the walls (determined by running a rigid wire through the tube), replacement is recommended to prevent selective removal or retardation of polar compounds. Check heater operation in these areas and make sure good thermocouple contact and placement is made.

Problems in the sample transfer lines such as deposits, leaks, and cold spots can result in many of the symptoms observed for inlet system and column malfunctioning (i.e.-less resolution, loss of sample, peak broadening, peak shifting, peak disappearance and ghost peaks). Leaks, of course, can be detected rapidly with soap solution and stopped; however, if the systematic elimination of trouble spots points to the sample transfer lines, they may be steam cleaned or removed and solvent cleaned. But in light of the time involved, replacement with new, clean tubing is usually the least time consuming and expensive.

In review, accurate troubleshooting in the column even system can save operating time. A systematic approach, using any combination of the hints and experiences listed in this chapter and in the chapter on Comprehensive Troubleshooting should help reduce down time, and allow the chromatographer to become more familiar with his particular GC unit.

Chapter 10

References

1. J.Q. Walker, Hyrocarbon Processing, 50, No.10, 99 (1971).

2. L.S. Ettre, "Open Tubular Columns in Gas Chromatography", 1st ed., Plenum Press, New York, 1965 (p. 79-82.)

3. Research Notes", published by Wilkens Instrument and Research (Now Varian-Aerograph), Fall Issue, 1961.

4. Private Communication, J.B. Maynard, 1971.

5. No. GC-AP-010, P10 Literature on Oven Designs for Gas Chromatographym 1967, Perkin-Elmer Corporation, Norwalk, Conn.

6. Private Communication, J.B. Maynard, 1971.

7. J.C. Giddings, Anal. Chem., 36, 1580 (1964).

8. D.J. McEwen, Anal. Chem., 35, 1636 (1963).

9. J.Q. Walker, Hydrocarbon Processing, 49,No 6, 178 (1970).

10. Private Communication, J.Q. Walker, 1970.

11. Perkin-Elmer Model 810 Operation Manual, Perkin-Elmer Corporation, Norwalk, Conn.

12. Hewlett-Packard Column Oven, 7620, Operation Manual, 1970 Hewlett Packard Corporation, Avondale, Pennsylvania.

13. Varian-Aerograph, Previews and Reviews, No. 5, Walnut Creek, California, 1967.

14. Private Communication, J.B. Maynard, 1971.

15. W.E. Harris and H.W. Habgood, Programmed Temperature Gas Chromatography, Wiley, New York, 1966.

Chapter 11

Gas Chromatography Detectors

As a sample component emerges from the column, the
time at which it is eluted and the quantity being
eluted is determined by the detector. The detector
then relays electrical signals, via an amplifier, to
the strip chart recorder establishing a permanent
record of the analysis. The arrangement of the detector in the GC system is shown in Figure 11-1.
Even though the primary function of the GC instrument
is separation, the detector provides the vital link
between the column separation and the chromatogram.
It is therefore important that the chromatographer
be well informed in the theory, operation, maintenance, and repair of detectors.

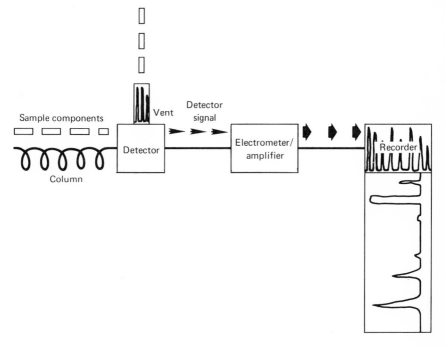

Figure 11-1   This Simple Diagram Shows the Position of
              the Detector in the GC System (1)

There have been more than twenty different types of
detectors developed for use in GC. However, only a
relatively few of these detectors are in common use
today. Most of these detectors and their uses will
be mentioned here, but only the more commonly used
detectors will be dealt with in detail.

Terms

No two detector types are operated identically, nor
are they all applicable to every type of detection
problem. However, some common terms are used to
describe detector performance, and these can serve
as a guide to selecting the proper detector and
evaluating detector performance.

Detectors are currently classified as belonging to
one of two categories according to how they measure
the sample. Detectors whose response is independent
of the flow rate of the carrier gas are called mass
flow rate detectors. While, in theory, these detectors are not affected by changes in carrier gas
flow rate, changes greater than about 25% will be
significant. The carrier gas flow rate must be held
constant for concentration dependent detectors.
With these detectors, the greater the amount of carrier gas eluted with the sample, the less the response from the detector.

Detectors can be categorized in terms of their selectivity. Some detectors will respond to almost any
type of sample and are called universal detectors.
Most detectors, however, will not respond to the
presence of certain groups of compounds. In fact,
some detectors are designed to respond specifically
to a small group of compounds such as phosphates or
halogens. The selectivity of a detector is illustrated in Figure 11-1. Column effluent from a sample
containing air, carbon dioxide, carbonyl sulfide,
hydrogen sulfide, propane, and propylene is split
1:1. Half of the effluent enters an electron capture
while the other half enters a flame ionization detector
(both are discussed in detail later.) Chromatograms obtained from these two detectors are vastly
different. The flame ionization detector registered
only the hydrocarbons, propane and propylene. The
electron capture detector did not respond to the
hydrocarbons at all, however, it did indicate the

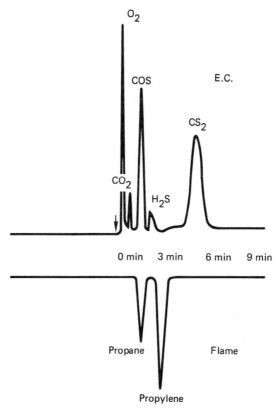

Figure 11-2   The Selectivity of Detectors is Illustra-
              ted by These Two Chromatograms Showing
              the Response of Two Different Detectors
              to the Same Mixture. (Courtesy Hydrocarbon
              Processing)(2)

presence of the other sample components.   Some detectors are highly selective and are usually directed toward a particular purpose.   Phosphorous detectors can be used, for example, to detect pesticides contained in the complicated mixtures resulting from the analysis of fruits.

Because the detector does not respond to a class of
compounds does not mean that the compounds will not have
some influence on detector performance.   For example,
while flame ionization detectors do not respond appreciably to water molecules, the presence of large

quantities of water in the sample can result in distorted peaks with some columns and lead to poor quantitation.

A quantity used commonly to describe detector performance is <u>sensitivity</u>. Essentially, sensitivity is a measure of the ability of a detector to signal the recorder of the presence of an emerging sample component.

Sensitivity measurements depend on the type of detector. The sensitivity of mass flow rate detectors is the peak area divided by the weight of the sample component as follows:

$$S_M = \frac{AS_RC}{W}$$

where,

S_M = Sensitivity of mass flow rate detector
A = Peak area (cm^2)
S_R = Sensitivity of recorder (mV/cm)
C = Reciprocal of chart speed ($MIN./cm$)
W = Weight of component (mg)

The sensitivity of concentration dependent detectors must account for the carrier gas flow rate. Sensitivity for this type of detector is given as follows:

$$S_C = \frac{AS_RC \ /u}{W}$$

where,

S_C = Sensitivity of a concentration dependent detector ($mV.CM^3/mg$)
A = Peak area (cm^2)
S_R = Recorder sensitivity (mV/cm)
C = Reciprocal of chart speed (min/CM)
$/u$ = Carrier gas flow rate at column outlet (cm^3/min)
W = Weight of component (mg)

Sensitivity is a convenient means of comparing detectors of the same type. It is difficult, however, to compare the sensitivity of a mass flow rate detector to the sensitivity of a concentration dependent detector. The compound whose peak is used to measure sensitivity is also significant; and, when comparing detectors, the same column and peak should be used to determine sensitivity. Obviously, the electrometer/amplifier range and attenuation

152

should be held constant also since this will directly influence the peak size.

The signal from the detector can be amplified by
nearly any amount. At high amplifications <u>noise</u>
will develop which is characterized by rapid fluctuation of the recorder pen resulting in a chromatogram of the type shown in Figure 11-3. The noise
(P) is measured over the maximum deflection of the
recorder pen when no sample is eluting. Since this
noise will make it difficult to accurately measure
small peaks, the <u>minimum detectable quantity</u> has
been defined as that quantity of sample producing a
peak with a height of 2P.

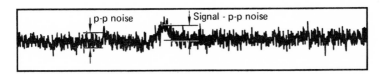

Figure 11-3   The Minimum Detectable Quantity is De
fined as a Peak which is Twice as High
as the Noise Level. (3) (Courtesy
Keithley Instruments Inc.)

<u>Linear Range</u> is important to accurate quantitative
determinations. If a detector responds to a given
amount of sample, twice as much sample should produce twice the detector response. All detectors
have a range of concentration over which this relationship will hold true. To measure the linearity
of a detector, the detector sensitivity is measured
over a wide range of sample concentrations. The
smallest concentration should be the minimum detectable quantity. If a semi-log plot is made of sensitivity as a function of the log of concentration, a
line will result which is straight over the linear
range. The upper limit of linearity is the point at
which the curve deviates from a line. The <u>linearity</u>
of the detector is then the upper limit of linearity
divided by the minimum detectable quantity. Because
a detector does not exhibit a wide linear range, this
does not mean that the detector is of little value.
It simply means that more calibration of peak area
versus concentration will be required for accurate
quantitation.

Kinds of Detectors

The detectors used in GC must now be considered separately. Some have different application, operate on
different principles, and require different precautions to be taken in their use.

Thermal Conductivity Detector

One of the oldest and still widely used type of
detector is the thermal conductivity detector (TCD)
or katharometer. A gas, flowing across a heated
filament, will cool that filament to some extent by
absorbing some of the heat. Different gases have
different abilities to cool the filament. If the
current heating the filament remains constant, and,
if a constant flow rate is maintained, the filament
will soon establish an equilibrium temperature.
This temperature can, in turn, be measured from the
resistance across the filament. Should the composition of the gas flowing across the filament be
changed, the ability of the gas to conduct heat away
from the filament will be altered. The change in
the temperature of the filament can again be measured by the filament resistance. This is the basic
principle behind TC detectors.

In Figure 11-4, a simple diagram of a TCD cell is
shown. The filament is contained in a heated block
through which the gas is flowing. The filaments
from the two cells are connected to form the elements
of a Wheatstone Bridge, as shown in Figure 11-5.
Through one cell, the reference cell, only carrier
gas is flowing. Column effluent flows through the
second cell. When no sample is eluting from the
column, the temperature, and thus resistance(20-40
Ohms)of each filament will be the same, balancing
the bridge. As sample, mixed with carrier gas,
enters the second cell, the filament temperature will
change resulting in an imbalance in the bridge circuit. The imbalance will, in turn, generate a signal which can be amplified and relayed to the recorder.

In practice, four filaments are used, two reference filaments and two sample filaments. These are

154

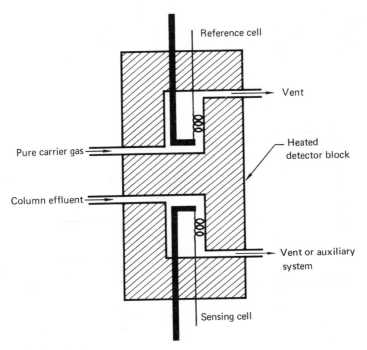

Figure 11-4 A Simple Two Cell Thermal Conductivity
             Detector (TCD) (4)

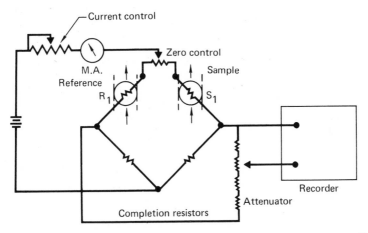

Figure 11-5 The Wheatstone Bridge Circuit for a Two-
             Cell TCD (5) (Courtesy J. of Gas
             Chromatog.)

arranged as seen in Figure 11-6. The advantage of this arrangement is that it provides twice the output signal and tends to stabilize the Wheatstone Bridge against outside temperature fluctuations.

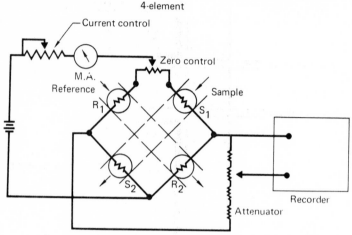

Figure 11-6 In practice, most modern TC Detectors have a Pair of Reference Cells and a Pair of Sensing Cells (5) (Courtesy J. of Gas Chromatog.)

Some consideration should be given as to the types of carrier gas to be used with TC detectors. Larger differences between the thermal conductivities of the carrier gas and the sample will create a more definite detector response. Since larger molecules exhibit lower thermal conductivities, a light carrier gas such as hydrogen or helium should provide the best results. The thermal conductivities of several common gases is listed in Table 11-1.

Other detector parameters should be considered to optimize TCD performance. An increase in filament temperature will increase the output signal. This will also result in a higher filament resistance bringing about a great increase in sensitivity. If the current is boosted too high, noise may develop and there is always the risk of burning out the filament. Detector block temperature is also important. It is heated primarily to minimize temperature fluctuations from the outside air and to prevent

Table 1.
Thermal Conductivities and Response Values
for Selected Compounds

Compound	TC at 100°C (Cal/cm sec deg x 105)	TC Relative to He = 100	RMR*
Carrier Gases			
Argon	5.2	12.5	----
Carbon Dioxide	5.3	12.7	----
Helium	41.6	100.0	----
Hydrogen	53.4	128.0	----
Nitrogen	7.5	18.0	----
Samples			
Ethane	7.3	17.5	51
n-Butane	5.6	13.5	85
n-Nonane	4.5	10.8	177
i-Butane	5.8	14.0	82
Cyclohexane	4.2	10.1	114
Benzene	4.1	9.9	100
Acetone	4.0	9.6	86
Ethanol	5.3	12.7	72
Chloroform	2.5	6.0	108
Methyliodide	1.9	4.6	96
Ethyl Acetate	4.1	9.9	111

*Relative molar response in helium. Standard:
benzene = 100. Taken from references 82 and 91. (5)
(Courtesy J. Gas Chromatog.)

condensation of sample vapors. Because sensitivity
will increase with a greater difference between
block temperature and filament temperature, it is
advantageous to keep the block temperature as low as
is practicle. High carrier gas flow rates would result in a lower filament temperature and a corresponding loss in sensitivity. For this reason, lower flow
rates are also beneficial.

Because the TCD is a concentration dependent detect-
or it requires good flow control. Fluctuations in
the flow rate of the carrier gas may result in a
noisy baseline or irregular drift of the baseline.
The TC cell can easily be checked for leaks by first
capping off the exit ports of the detector cells.
Then, pressurize the detector cells to 30-40 psig
and turn off the pressure source. If the pressure
remains constant, the cells are leak tight. This
same method is useful for leak checking the entire
flow system, though small leaks may not be immediately obvious. Extreme changes in the carrier
gas flow caused by loose particles trapped in the
gas line (a crumbling injection port septum is a
typical source) can result in difficulties in zeroing the recorder pen or the pen may go completely
off scale and not return.

When operating at carrier gas flow rates of 10 cm^3/min
or less, it is advisable to install about three to
six inches of 0.01 inch i.d. tubing at the detector
outlet. This tubing will act as a restrictor to
prevent back diffusion of atmospheric air into the
detector cell. A similar restrictor is advisable
at the detector inlet when backflushing or multi-
column valves are used as part of the GC system. A
restrictor in this position will dampen pressure
surges into the cell and result in better cell stability.

Filaments in the TC cell require some care. The
filament current should be monitored; and, if the
current does not remain constant, baseline noise or
a gradually drifting baseline will result. If there
is no current showing on the meter, the problem is
either a meter malfunction or a broken filament. A
broken filament will result in the recorder pen remaining off scale. Filaments can burn out, particularly if highly reactive samples have been run

through the detector. To protect the filaments, it is advisable that the filament current knobs be turned down as low as possible before turning the cell on. This reduces the power surge to the filament resulting in a longer filament life. Another precaution which will increase filament life is to be sure that the detector cell has been purged with inert carrier gas before turning the cell current on. This can prevent oxidation of the filaments. Problems which appear to be caused by the filament may only be the result of loose or faulty connections between the detector and power supply. (Detector Cleaning-AppendixA)

Flame Ionization Detector

Even though a more recent development than the thermal conductivity detector, the flame ionization detector (FID) is being used more today than any other type of GC detector. The FID is nearly a universal detector responding to all but a few gases, such as the so called "permanent" gases, nitrogen, oxides of nitrogen, H_2S, SO_2, COS, CS_2, CO, CO_2, H_2O, and HCOOH. Actually, this insensitivity can be used to good advantage by using H_2O or CS_2 as a solvent, since large solvent peaks are eliminated. The presence of some of these compounds can have some effect on detector performance however. While other detectors are more universal, the FID has the advantage of being able to detect concentrations as low as 10^{-12} gm/ml.

Linearity of response is also a valuable asset of the FID. While a linearity in the range of 10^4 is not uncommon in detectors, the linearity of an FID has been shown to cover a range of 10^7 (8).

Even though flame ionization detectors are extremely sensitive, they are not particularly complicated. The basic design of the detector is shown in Figure 11-7 (9). An electrical field is established between a negatively charged hydrogen burner jet and a positively charged electrode. Column effluent is ionized in the flame and the negative ions are collected at the detector anode. To insure total ionization of the effluent, an excess of oxygen is introduced into the FID chamber by introducing air at the base of the flame jet.

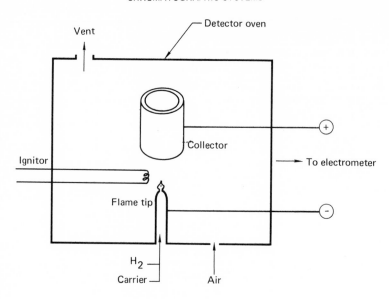

Figure 11-7   The Components of a Flame Ionization
              Detector (FID) (9) (Courtesy Anal.
              Chem.)

While a given set of detector conditions can be used
for a wide range of GC applications, the detector
should be optimized for each column, column flow rate,
and sample. A plot of sensitivity vs hydrogen flow
rate as shown in Figure 11-8 (10), will show a maximum sensitivity probably near 30 ml/min. Changes in
carrier gas flow rate of not more than 25% should
not greatly affect the flame response (11). The flow
of air into the detector should be of the order of
10 times the flow rate of H_2. Various other performance factors for FID, such as flame jet temperature
and applied voltage are discussed by McWilliam (12).

The FID is subject to some limitations. Since column
effluent is destroyed during detection, use of the
effluent for further analysis necessitates the use
of a effluent   splitter. Response of the detector to pure hydrocarbons (containing only hydrogen

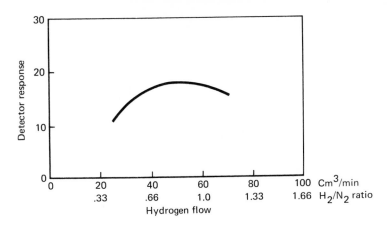

Figure 11-8   The Sensitivity of an FID is Affected by
              the H_2 Flow Rate of the Flame (10)
              (Courtesy Hewlett Packard)

and carbon) is approximately linear with respect to
carbon number; substituted groups, however, will
decrease the sensitivity of the detector. For substituted hydrocarbons (substituted with atoms of
halogen, sulfur, phosphorous, oxygen, or nitrogen, (for
example the FID requires the use of calibration
curves for accurate quantitation.

An FID developed by researchers at Shell Oil Company
(13) provides approximately 200 times the useable
sensitivity of conventional FID apparatus. This is
accomplished through changes in the detector geometry
and in the applied voltages. An excellent review
of factors related to TCD and FID operation has been
published by Dietz (14).

Flame detectors are very reliable instruments, however, they too require periodic maintenance. It may
be found, for instance, that the ignitor fails to
ignite the flame. This can be due to a broken ignitor coil or lack of voltage from the electrometer.
Ignitor failure can usually be traced, though, to
a defective ignitor voltage cable. Heat at the
junction of the detector and ignitor cable can cause
an oxidative film to build up, thus preventing a
good electrical connection. The junction can easily

be cleaned with a piece of emery cloth or other abrasive. Before igniting the flame, always be sure that the detector temperature is at least 125°C to prevent condensation of water formed from the combustion of hydrogen. Never attempt to light the flame with an ordinary match. Phosphorous and other foreign matter could fall into the detector giving rise to noise. If it is necessary to ignite the flame without the use of the igniter, construct a small wick by inserting a pipe cleaner into a pyrex glass tube. Dip the pipe cleaner in methanol and ignite it. This flame can be used to safely light the FID.
If the flame ignites but fails to remain lit, check first to see that hydrogen flow and air flow are at their proper levels. If all flow controllers and pressure regulators are in their proper position, check to see if there is a flow coming from the flame tip by connecting the flame tip to a flow meter through a section of teflon tubing (to prevent contamination of the tip). A restriction of the flame tip can sometimes be removed with a piece of thin clean wire. Caution: Before handling the detector in any way be sure to turn off all power sources. A high carrier gas flow rate may cause the flame to blow out. A splitter can be used to decrease the amount of carrier gas reaching the detector if the high flow rates are desirable. Large samples entering the detector can overload it and quench the flame.

It is not difficult to determine whether the source of the noise is from the detector or if the source is elsewhere. First, remove the signal output and cover the input connector with a piece of aluminum foil. If the baseline is still noisy, the problem lies in the electrometer or recorder, not the detector. If a steady baseline results, the source of the noise is probably the detector. Inspect the entire detector for build up of soot or other extraneous particles and clean it if necessary (see appendix A). Excessive column bleed which has resulted from poor column or excessive column temperatures can usually be detected by examining the flame. Impurities entering the flame will give the flame color while a normal flame is invisible. Be sure that a pure carrier gas is used and that air entering the detector is filtered as these are both causes of detector noise. If the ignitor coil is improperly positioned above the flame tip, a great deal of noise

162

will result. Since detector geometries vary, consult the instrument manual for proper positioning. A
frequent cause of detector contamination is column
bleed. If the teflon insulators have become encrusted
with deposits, this can be alleviated by scraping them
with a sharp knife. Reducing the column operating
temperature is the best way to reduce column bleed
entering the detector.
An alternate approach is to adapt an adsorbing column
between the separation column and the detector. This
adsorbing column can be made by filling a short 6
inch by 1/8 inch O.D. section of tubing with a 3%
solution of CW20M-TPA (Carbowax 20,000-Terephthalac
Acid) liquid phase on a solid support. This column
will absorb column bleed for a time, but will not
function indefinitely. However, several columns of
this type can be prepared simultaneously.

The detector may become contaminated to the extent
that it cannot be cleaned simply by raising the detector temperature. It will then be necessary to
clean the detector as outlined in appendix B.

Electron Capture Detectors

The electron capture detector (ECD) was developed
around 1960 by Lovelock and Lipsky (15). Another
type of ionization detector, the ECD, uses a radioactive source to generate the ions measured in these
detectors. Rapidly moving free electrons from the
radioactive source, (H^3, Ni ^{63}, or Kr^{85}) will be
captured by molecules in the detector to form stable
negative ions or carged atoms. This reaction occurs
by one of two methods:

$$a) \quad AB + e^- \; ---\!\!\gg (AB)^- + energy$$
$$b) \quad AB + e^- \; ---\!\!\gg A^- + B \pm energy$$

Not all molecules exhibit the same capability to
capture the free electrons. This is a measure of
the electron affinity of a molecule. Those molecules
showing lower capability for capturing free electrons
are said to possess a lower electron affinity.

In the EC detector, a carrier gas, usually 95% argon
and 5% CH_4, is flowing past an ion source as shown
in Figure 11-9. The free electrons ionize the nitrogen creating a current flow. As samples with

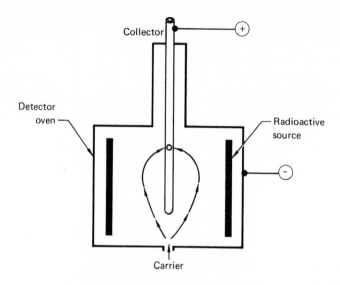

Figure 11-9   The Components of an Electron Capture
              Detector (ECD) (11) (Courtesy Anal.
              Chem.)

higher electron affinities enter the detector chamber, the current flow is reduced. Unfortunately,
the amount of current reduction is not only a function of the amount of sample present, but also the
electron affinity of the sample. For quantitation
then, a calibration must be made separately for
each sample component.

The ECD exhibits considerable selectivity. It is
highly sensitive to halogenated compounds (I, Br, Cl,
F) nitrates, conjugated carbonyls, and certain organo
metallic compounds. It is not sensitive, however,
to compounds with low electron affinities, such as
aliphatics and alcohols. The selectivity of the ECD
for halogen compounds has lead to its extensive use
in the field of pesticide research. For compounds
such as lindane and aldrin ECD can detect quantities
as low as 0.1 picogram.

The primary source of problems with EC detectors is
their low range of linearity. While FID detectors
show linearities over a range of 10^7, an ECD cannot

exceed 5×10^2. Linearity will vary according to
the radioactive source used. Tritium, H^3, provides
the widest linear range as seen in Figure 11-10.
Ni^{63} provides a linear range of only about 50, but
the linearity does not level off sharply as with H^3.
Use of a calibration curve can extend the useful
range of ECD with an Ni^{63} source to nearly 10^3.

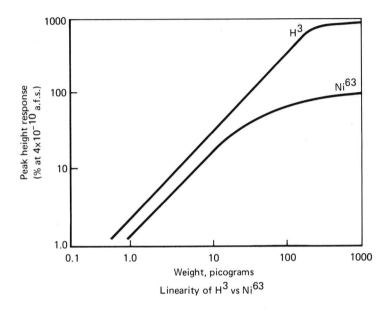

Linearity of H^3 vs Ni^{63}

Figure 11-10   A H^3 Source for an EC Detector exhibits
                a Wider Linear Range than does the
                Ni^{63} Source (17)   (Courtesy Varian
                Aerograph)

Other factors should be considered in the choice of
the radiation source. H^3 is the least temperature
stable of the two isotopes and is limited to 220° C
by the Atomic Energy Commission. Low vapor pressure
ples can condense on the tritium (H^3) detector causing harmful contamination. Ni^{63} sources are capable
of withstanding temperatures up to 350°C lessening
the danger of contamination of the detector and reducing the danger to the operator of harmful radiation. Recently a tritium source has been described
with a maximum operating temperature of 300°C (17).

This will undoubtedly increase the popularity of tritium sources for ECD detectors.

Many studies have been made concerning optimizing of ECD operation (18,19). Detector temperature, applied voltage, and carrier gas flow rate will influence sensitivity. A temperature increase will bring about an increase in sensitivity, but an increase in carrier gas flow rate will result in a decrease in sensitivity. Variation in applied voltage also influences sensitivity and can, if desired, contribute to the selectivity of the detector. Since increasing the applied voltage will decrease sensitivity, compounds with low electron affinities can be screened out by a high applied voltage. A pulsed, rather than a continuous voltage is often used to minimize the effects of flow rate and temperature changes.

If the standing current of the EC detector is low, or, if the baseline dips negative after peaks, cleaning is necessary. Cleaning requires proper AEC authorization, though the method is simple as outlined in Appendix B.

The ECD requires accurate temperature control. If the temperature is allowed to drift, so will the baseline. AEC regulations require that the unit be equipped with upper temperature limit controls to prevent the detector from rising above the limit for the radioactive source. An exhaust vent above the ECD is also required in the event that the controls do not function properly.

Except for cleaning procedures and detector temperature control requirements, troubleshooting methods for the ECD are similar to those for the FID.

Some problems are involved in the operation of an ECD. Quantitation requires extensive calibration and usually it is difficult to obtain good accuracy over a wide range. Care must be taken to avoid contamination of the detector and a hot septum should be avoided. The ECD is highly sensitive to O_2 (1 ppm) and care must be taken to eliminate any possible air leaks in GC system. Some methods are available for removing small amounts of oxygen and water from the carrier gas stream before it enters the detector. One of these, called the Oxisorb (TM) vessel (manufactured by Analabs, Inc., North Haven, Conn.) can hold oxygen impurities below 0.1 ppm (where inlet does not exceed 15 ppm) and can hold water below 0.5 ppm (where inlet does not exceed 10 ppm). Additionally, "wet" samples should be dried in a suitable fashion prior to analysis. Even traces of water in samples cause appreciable noise problems which last for some duration due to adsorption effects within the detector.

Alkali Flame Ionization Detectors

Phosphorous containing compounds are selectively detected by the alkali flame ionization detector. Similar in operation to the FID discussed earlier, the major difference is the addition of a quantity of an alkali salt placed near the tip of the flame, as seen in Figure 11-11 (20). Cesium bromide

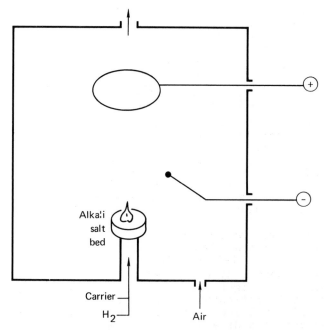

Figure 11-11   The Basic Parts of an Alkali Flame Ionization Detector (AFID) (20) (Courtesy Anal. Chem.)

(CsBr) and rubidium sulfate (Rb_2So_4) have both
been used as the salt in a variety of configurations
including pellets, impregnated screen and coated
wire loops. A lower air flow rate is also used.
Generally, an air flow rate of approximately 130 ml/min is
used as opposed to flow rates of 300ml/min or more
for normal FID. It is essestial to maintain constant
hydrogen and air flow rates for good quantitation.
This demands the use of good flow controllers. The
stability of hydrogen flow cannot be overstressed.
Adjustment in the flow rates can optimize this detector for nitrogen compounds. While no clear theoretical explanation for the operation of the AFID
exists at this time, it nonetheless enhances the
response to phosphorous compounds by 5000 to 1 and
the response of nitrogen compounds by 25 to 1 over
the response for pure hydrocarbons.

Some precautions should be exercised in the operation
of an AFID. Use of more than 1:1 ratio to sample of
solvent will seriously upset the detector response.
A considerable amount of time is sometimes necessary
for the detector to return to normal. The sensitivity of the detector is subject to gradual changes
because of losses from the alkali pellet, which
must be replaced periodically. It is therefore,
necessary to make frequent calibrations of the detector to maintain accurate quantitative results.

Flame Photometric Detectors

The increasing interest in air pollution has added
to the popularity of another type of detector.
Flame photometric detectors (FPD) can be adjusted
to obtain selectivity for either sulfur or phosphorus
compounds, both of which are common constituents of
air pollutants. Not only are these detectors highly
selective, they are also sensitive enough to detect
nanogram quantities of organo-phosphorous compounds.

The basic idea behind the FPD is the measurement of
emittance from the light of a hydrogen flame. As
shown in Figure 11-12 carrier gas mixed with oxygen
rich air enters a hydrogen filled chamber through
a burner tip. Light from the flame impinges upon a
mirror and is reflected to an optical filter. The
optical filter allows only light of either 526 mµ
(for phosphorous detectors) or 394mµ (for sulfur

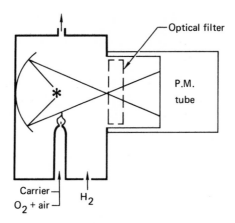

Figure 11-12   The Basic Parts of a Flame Pnotometric
Detector (FPD)   (20)   (Courtesy Anal.
Chem.)

detectors) to pass through to a photomultiplier tube.
Current from the photomultplier tube is sent to the
electrometer.

Aside from its specificity, other factors keep the FPD
from more widespread use. First, the FPD exhibits
little or no linearity. For accurate quantitation,
a calibration curve must be determined for each compound in the sample. Due to the arrangement of the
flame and the large amount of hydrogen present,
sample solvents are likely to extinguish the flame.
In more recent models a manually operated column effluent stream diverter is available to prevent solvent
from entering the detector. (The new Bendix model)
Another universal detector is the cross section detector developed originally by Deisler and co-workers
around 1955 (21). While lacking something in sensitivity, these detectors are nonetheless dependable
and relatively simple. Beta particles from a
tritium source ionize samples entering the detector.
A carrier gas is used which has a low cross section
(H_2 or He + 3% CH_4) thus diminishing the probability
of an ionizing beta particle collision. A potential
applied across two closely spaced electrodes will
collect any ions formed from beta particle collision.
When a sample enters the detector which has a higher
cross section than the carrier gas, more collisions

will result, producing a higher current. This current increase is amplified and recorded. A typical cross section detector is shown in Figure 11-13 (22).

The sensitivity of this detector varies somewhat with the cross section of the molecule being detected. Clark (23) showed that the sensitivity could be predicted on the basis of their calculated molecular cross sections. Once again, AEC regulations must be adhered to in the handling of the radioactive source.

Helium Detector

Helium detectors (HeD) provide for the detection of minute quantities of permanent gases. These de-

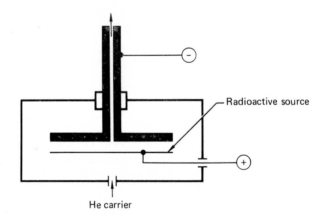

He carrier

Figure 11-13   The basic parts of a cross-section detector and a helium inoization detector (HED; the chief difference between the two being the use of the field gradient with the HED (22) (Courtesy Anal. Chem.)

tectors are physically very similar to the cross section detector. The major difference is the utilization of a very high field gradient across the detector electrodes. This causes the He carrier gas to enter a metastable He* state having an ionization potential of 19.8 eV. Only neon, among the permanent gases, exceeds this ionization potential, thus all samples are ionized producing a positive signal.

HeD must be used only with gas solid chromatography, since column bleed from liquid loaded columns will quench the metastable He* before it can ionize sample molecules. Quantitatition in the ppb range is difficult with the HeD and somewhat limited by the lack of an adequate calibration system. Still, HeD is fifty times more sensitive to methane, for example, than the FID. This in itself will harbor to the continued use of this detector.

The Gas Density Balance

One of the original GC detectors was the gas density balance (GDB). It is not in use much today, but it still has some properties which make it useful. Foremost of the advantages of a gas density balance is that the column effluent never comes into contact with the detector sensors. This makes the GDB useful for detection of corrosive materials which otherwise might cause damage to other detectors. In addition to this, the GDB is comparatively simple, requires no calibration, can be used with readily available carrier gases (nitrogen, argon, carbon dioxide), Sulfur Hexaflouride and is non-destructive.

Operation of the GDB can best be described by referring to the diagrams in Figure 11-14 (4). In Figure 11-14a, a reference gas identical to the column carrier gas enters at C. Column carrier gas enters at D. When the gas entering at C is identical to the gas entering at D (i.e., no sample in carrier gas stream), flow past sensors A and B will be identical. Sensors at A and B are heated elements connected through a Wheatstone Bridge.

If, as in Figure 11-14b, the column effluent contains a sample more dense than the carrier gas, downward flow of the sample will retard flow of the reference gas past sensor B. The result is a change in the amount of heat carried away from sensor B causing a change in the element resistance with subsequent imbalance of the Wheatstone Bridge.

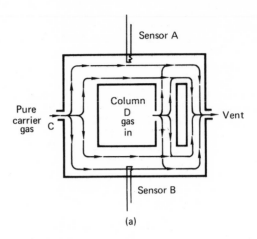

(a)

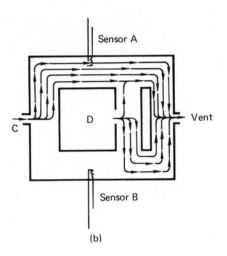

(b)

Figure 11-14   The principles of operation of a gas density balance: (a) no sample from the column, equal flow across sensors A and B, (b) with sample from the column, the flow across sensor B is blocked off forcing more flow through sensor A. (25)

A few considerations should be made when operating a GDB. Sensitivity is improved with a greater difference between the density of the carrier gas and the density of the sample. Hydrogen and helium, however, are prone to diffuse rather easily and nitrogen is generally preferred as the carrier gas. Column flow rate is not critical with the GDB, however, the reference gas should have at least a 20 ml/min faster flow rate than the column carrier gas to prevent backflow.

With the GDB, it is possible to make accurate quantitative analysis without the tedious calibrations necessary with many types of detectors. Detector response is easily converted to weight per cent by multiplying peak area by a factor related to the molecular weights of solute and carrier gas. The principle is presently being used in the Mass Chromatograph.

$$\text{wt. factor} = \frac{A(MW)s}{(MW)_s - (MW)_c}$$

where
A= Area of peak
$(MW)_s$ = Molecular wt of solute
$(MW)_c$ = Molecular wt of carrier gas

$$\% \text{ wt of } s = \frac{\text{wt factor of s}}{\text{all wt factors}} 100$$

Summary

Because of its versatility and high sensitivity, FID is the most widely used detector today. Close behind is the TCD which is one of the early detectors used in GC. In the future it is probable that more work will be done on the use of specrographic methods of detection. Currently, the major problems in this area lie in interfacing techniques and the high cost of infrared and mass spectrographic equipment. Optimum utilization of current detector technology, however, can provide a great deal of accurate and meaningful information. Table 11-2 summarizes the characteristics of several GC detectors. Detector Cleaning is described in Appendix "A".

173

Table 11-2 Comparison of Parameters of GC Detectors

Detector Parameter	Thermal Conductivity	Flame Ionization	Alkali Flame Ionization	Electron Capture	Cross Section	Gas Density	Flame Photometric	Helium Ionization
Linear Dynamic Range	10^4 to 10^5	10^6	10^3	10^3	3×10^5	10^5	500	5×10^3
Minimum Detectable Quantity, gm/sec	10^{-9}	3×10^{-12}	10^{-11}	3×10^{-14}	3×10^{-11}	10^{-10}	2×10^{-12}	4×10^{-14}
Minimum Detectable Conc., Parts by Vol	10^{-5}	10^{-11}	10^{-9}	10^{-13}	2×10^{-6}	10^{-6}	4×10^{-10}	10^{-13}
Carrier Gas	H_2 & He	He, N_2	He, N_2	Ar,CH_4 & N_2	H_2 & $5\%CH_4$	SF_6	N_2	Pure He
Detector Vol, ml	10^{-2} to 1.0	Ca. 10^{-4}	Ca 10^{-3}	0.8	8×10^{-3}	0.3	ca. 10^{-3}	0.3
Applicability to Column Sizes	Packed & Capillary	Packed & Capillary	Packed & Capillary	Packed	Packed & Capillary	Packed	Packed & Capillary	Packed
Applicability to Gases & Liquids	All	Organic Compounds	Phosphorous & Halogen Cpds	Oxygen & Halogens	All	All	Sulfur & Phosphorous	All
Utility in Trace Analysis	Low to Medium	Extremely High	Extremely High	Extremely High	Medium	Medium	Extremely High	Extremely High
References	5	12	16	19,20	21	22	23,24	22

Chapter 11

References

1. J. MacDonald, McDonnell Douglas Corp., Private Communication, 1971.

2. J.Q. Walker, Petroleum Refiner, 44, 34 (1965).

3. Keithley Instrument, Inc. G.C. Technical Bulletin, p. 2, 1967.

4. J. MacDonald, McDonnell Douglas Corp., Private Communication, 1971.

5. A.E. Lawson, Jr., and J.M. Miller, J. of Gas Chromatography, 4, 273, (1966).

6. Ibid. 4, 274, 1966.

7. Ibid. 4, 276, (1966).

8. L. Ongkihong, in Gas Chromatography 1960,R.P.W. Scott, Ed., Butterworths, London(1960) p.46.

9. C.H. Hartman, Anal. Chem, 43, 113 (1971).

10. F & M Scientific Corp., Pre-print No. 1, Pittsburg Conference of the ACS, 1966.

11. C.H. Hartman, Anal. Chem., 43, 114 (1971).

12. I.G. McWilliam, J. Chromatography 51, 391 (1970).

13. B.O. Prescott, H.L. Wise, and D.A. Chestnut, U.S. Patent No. 3, 451, 780 (1969).

14. W.H. Dietz, J. Gas Chromatography, 5, 68 (1967).

15. J.E. Lovelock and S.R. Lidsky, J. American Chemistry Society, 82, 431 (1960).

16. C.H. Hartman, Anal. Chem., 43 ,117, (1971).

17. C.H. Hartman, D. Oaks, T. Burroughs, Technical
     Bulletin 130-66 ,Varian Aerograph.

18. D.C. Fenimore and C.M. Davis, in "Advances in
     Chromatography", Ed. A. Zlatkis U. of Houston,
     Houston, Texas. 1970, p. 130.
19. M. Scolnick, Technical Bulletin 132-67, Varian-
     Aerograph, 1967.

20. C.H. Hartman, Anal. Chem., 43, 118, (1971).

21. P.F. Deisler, K.W. McHenry, Jr., and R.H. Wilhelm,
     Anal. Chem, 27, 1366 (1955).

22. C.H. Hartman, Anal. Chem., 43, 120, (1971).

23. S.J. Clark, Gas Pipe, No. 1, Jan. 1963, Jarrell-
     Ash Co., Waltham, Mass.

24. A.J. Martin, Technical Bulletin, "Mass Chroma-
     tograph", Chemalytics Corp.,1971.

25. J. MacDonald, McDonnell-Douglas Corp., Private
     Communication, 1972.

Chapter 12

Integral Electronics

One of the aspects of GC instrumentation which is best understood in both a qualitative and quantitative sense is the electronics section of the instrument. With only the most fundamental knowledge of electronics, the gas chromatographer can understand the operation of at least two of the basic components present in many gas chromatographs, i.e. the electrometer and recorder. With the advent of capillary columns, more sensitive detectors were required because of the small sample sizes which must be injected into the columns. In addition the use of gas chromatography for analysis of trace components requires an ionization detector(1). The use of flame ionization (FID), argon ionization, electron capture, and cross-section detectors requires an electrometer amplifier for presentation of the detector output on a strip chart recorder. The quantitative analysis of multi-component blends could require much longer analysis times if it were not for automatic digital integrators. While the use of the ionization detectors and automatic digital integrators is by no means limited to analyses with capillary columns, it is with this type of analysis that this combination is most useful.

This chapter is intended to cover the basic operating theory and maintenance techniques of T.C. power supplies, electrometers, recorders, integrators, and describe various kinds   data-handling techniques. Since the material is intended for the practicing gas chromatographer, only fundamental theory will be covered. Operating procedures for electrometers, recorders, and integrators will be covered, along with potential sources of trouble and possible corrections. With only a fundamental knowledge of the operation of these systems, diagnosis of many of the electronic problems can be simplified.

Electrometers

The first link in most GC integral electronic systems is
the electrometer. The signal output from ionization
detectors, whether it be flame, argon, electron capture, or cross section is an extremely small electrical current, i.e. a flow of electrons. If this
small current flows through a high resistance element, a significant voltage is developed across the
resistor. The current, resistance, and voltage are
related by Ohm's law:
$$E = I \times R,$$
where E is the voltage, I is the current, and R is
the resistance in units of volts, amperes, and Ohms,
respectively. If a minute electron current of 10^{-11}
amps flows through a resistance of 10^{10} Ohms, the
voltage across the resister must be 0.1 volt or 100
millivolts, as given by Ohm's law. This is more than
enough voltage to drive a 1 mV.potentiometric recorder full-scale. However, potentiometric recorders
can only accept signals across a resistance of several thousand ohms. It is the function of the electrometer-amplifier to develop an identical voltage
across the relatively low resistance elements of the
attenuator. A potentiometric recorder (to be discussed later) can then be connected across the attenuator
for graphic presentation of the detector output
signal.

Figure 12-1 is a block diagram of a typical ionization detector tube-type electrometer hook-up. For

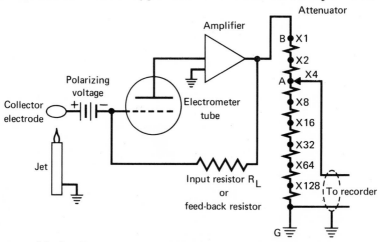

Figure 12-1 Basic Electrometer Configuration
(Courtesy Varian Aerograph) (3)

purposes of illustration the FID has been chosen; however, the same principles would apply to other ionization detectors. An operational amplifier is connected across the output of the electrometer tube. As the electrons flow into the grid, this pushes the grid to a more negative potential. In doing so the plate potential becomes more positive (a fundamental characteristics of triodes) and a greater demand for electrons is created. The plate current must come from the grid and hence the grid potential would be swung more positive if electrons were not supplied to it from the amplifier. In order to supply these electrons, the amplifier must develop a potential (voltage) across the range (or input) resister to "pump" electrons to the grid. The amplifier voltage output signal (V_a), input resister (R_L), and "feedback current" are related by Ohm's Law:

$$V_a = i \times R_L .$$

The output voltage (V_a) is impressed across a several thousand ohm resistance attenuator. The voltage drop across the resistance elements of the attenuator circuit can be "picked off" at points corresponding to the various stages of attenuation.

The amplifier, as mentioned, must develop a voltage, hence a negative feedback current to the input circuit is needed. The use of the negative feedback current is necessary for three reasons:

a.) To reduce the time constant of the electrometer. The time constant of the input circuit is the lag in response between the input and the output signals. It is determined by the product of the input resistance, which may be 10^{11} ohms, and the capacitance of the input cable from the detector to the electrometer. The capacitance may be 5-25 $\mu\mu f$ per inches of cable. The product R x C, in units of seconds could be significant and could affect the response of the system. The effect of the negative feedback is to reduce the time constant by the gain of the amplifier. The gain may be as high as about 10^4 .

Time constant (secs.) = $\dfrac{\text{Input resistance (ohms) x Capacitance (farads)}}{\text{Amplifier gain}}$

In this way the time constant is minimized.

b.   To operate within the linear region of the electrometer tube.   If the ionization current from the flame
were allowed to drive the grid potential up and down
without limit, operation of the tube could move into
a region of non-linear response. This non-linearity
would also be reflected the amplifier output to the
recorder. In addition the saturation point of the
tube could be reached; beyond this point any increase
in the input signal would not be reflected in the
output signal. The use of negative feedback maintains operation of the electrometer tube in the linear
region of the curve.

c.   To maintain a constant detector jet bias potential. The polarizing voltage of the collector electrode is the voltage of the battery minus the voltage
drop across the input resister, a variable quantity.
Each time a component passes through the flame the
jet bias would change and non-linear operation would
result in feedback were not used.

The input resistor and the attenuator network are the
means by which we vary the sensitivity range of the
electrometer. The input resistors are in decade steps
and the attenuation is normally in binary steps. The
attenuator can also be separate from the electrometer, in series with the recorder. The higher the
resistance of the input resistor (load resistor,
feedback resistor), the greater will be the voltage
which the amplifier must develop in order to drive a
given current to the electrometer tube grid. The
glass envelope of the resistor does not have an infinite resistance and thus sets a limit to which the
resistor can be pushed. If the envelope gives less
resistance to flow than the resistor wire, the electrons would flow through the path. The grease from a
fingerprint on an input resistor can provide a path
of less resistance to electron flow. For this reason
all parts of the input circuit, especially the input
resistors, should not be handled by the glass envelope.
If they must be handled, use the external leads and
preferably, with a pair of pliers. Accidental paths
for leakage can lead to a drastic loss in sensitivity.
The electrometers used in gas chromatographs have
various means to indicate the level of sensitivity
or the input resistor selected. For example, Hewlett
Packard uses ranges 1, 10, 10^2 , 10^3 , and 10^4, with
range 1 being the most sensitive (2). Varian-Aerograph uses ranges 0.1, 1, 10, and 100.(3) The most

sensitive range is the one which includes the highest input resistance, as is needed to produce measurable IR drops from the very small currents produced by ionization detectors. The attenuator is the second means by which we change sensitivity (sometimes an integral part of the electrometer, sometimes separate). The attenuator range may cover a span of from x1 (most sensitive) to x 1024, in binary steps. Some instruments have a ternary attenuator (x1, x3, x10, x30, etc.). The attenuator is the means by which we ordinarily reduce sensitivity to the recorder during the course of a run. Generally the attenuator is used only to reduce output to the recorder and keep it on scale. Most electrometers can handle the input from GC analysis on one current range without saturation due to a wide dynamic range.

Figure 12-2 is a block diagram for a new solid state

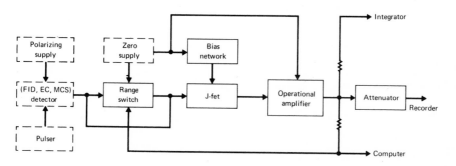

Note: Dotted blocks not part of electrometer assembly.

Figure 12-2 Solid-State Electrometer Functional
(Courtesy Hewlett-Packard) (2)

electrometer available from Hewlett-Packard, which
operates in essentially the same manner only using
solid-state components instead of tubes (2).

Functional Operation of the Electrometer

There will normally be eight different control knobs
on the electrometer. They are: a) mode control, b)
detector input, c) balance, d) polarity, e) range,
f) coarse, g) fine, and h) attenuation(if attached).
Many electrometers will also have a control for a
D.C. pulser built into the system. In the use of the
electron capture detectors most workers have found
that a pulsed D.C. potential on the collector electrode is better than a constant D.C. potential; hence,
this section will be used only if the electron capture detector is used. In beginning the operation
of the electrometer the first step is to turn the
mode control to the warm-up position. In this position voltage is applied to the filament, but no
potential is applied to the grid or plate. A warm-
up period of 15-60 minutes is usually required.
Under normal operating conditions the instrument
should be left in the warm-up position, but should
be turned off for long periods of nonuse. After the
recommended warm-up period, the mode control is turned
to the appropriate position corresponding to the
ionization detector being used.

The first point to be considered under "functional
operation" is adjustment of recorder zero. Although
this is independent of electrometer operation, this
step is performed first, with the electrometer disconnected from the recorder(or at infinite attenuation).
Recorder zero means that the recorder pen is at
zero while the recorder is operational, but no input
signal is applied to it. This adjustment is necessary so that signal attenuations will be linear.
To insure that no signal is getting to the recorder
the attenuator is indexed to infinity or the leads
to the recorder are removed. The recorder is zeroed
by the zero control on the recorder amplifier. For
example, the Honeywell Model 15 has a black dial in
the center of the amplifier section for this adjustment. The Honeywell Model 16 control is on the
right side along with the span, gain, and damping
controls. Rotation of this control moves the pen
into coincidence with the zero of the chart paper

and scale. The recorder zero drift is usually not large and this adjustment may need to be performed only weekly (or less often). This is also a good time to determine whether the scale zero and chart paper zero are in alignment. The next step is to balance the electrometer. The electrometer output is fed to the recorder, with the detector disconnected from the electrometer. With the attenuator at x1, the balance control is turned until the recorder pen is again at zero. In essence we are balancing the two matched electrometer tubes so that there is no output to the recorder in the absence of a signal input to the electrometer. If the instrument being used is equipped with two or more detectors (dual flame, flame-electron capture, etc.) the detector input selector switch is now turned to feed the desired signal into the electrometer. The instrument is now switched to the desired range. For a flame ionization detector with a small bore capillary column and a sample size of, for example, 0.5-1.0 μl split 100/1, a range of 10^9 ampers is a good one to start with. For a packed column and a sample size up to 10 μl with no split, a range 100 to 1000 times less sensitive is needed. The coarse and fine zero controls are now used to again bring the pen to zero, bucking out any background signal (these controls are sometimes labeled "bucking voltage"). The polarity switch is used to obtain the proper polarity on the bucking voltage. The buck-out operation is ordinarily performed at an attenuation more sensitive than will be used for the analysis.

Problems and Solutions

Following are some of the symptoms that may be observed when the electrometer malfunctions. Since the detector-electrometer-recorder combination produces the chromatogram, the observed symptom may be in any one of the three components and isolation of the trouble is necessary. By electrically isolating each component the trouble can be located. A weak signal response may simply be due to selection of the wrong input resister setting or attenuation. It may also be due to a shorted input resistor due to dirt or grease on the resistor envelope as mentioned previously. If it is impossible to balance the electrometer in the absence of a signal from the detector (input cable disconnected or range selector switch

in the balance position) the trouble probably lies
in the electrometer tubes and they should be replaced.

Loss of signal may also be due to weak batteries in
the collector electrode biasing circuit. This voltage is usually kept from 300-350 volts, and weak
batteries will not produce efficient collection of
electrons.

If the recorder baseline is stable without electrometer output, reconnect the electrometer to the recorder (or change to an attenuation other than infinity) and observe the recorder. If noise is observed at this time the problem must be in the electrometer (if no input current is coming from the detector). If no noise is observed, disconnect the
electrometer input cable at the detector, cover the
end with foil to eliminate pick-up stray signals,
and observe the recorder. Noise can be produced if
the input cable is not properly shielded and grounded. The magnetic field associated with alternating
current can act like a generator and cause electron
flow in the input circuit. Most input cables are
provided with a braided shield which is attached to
a grounding point. Any current which is generated
by an external alternating current or changing magnetic field will flow in the shield to ground and will
not show up in the output. If no noise has been
observed to this point, the problem lies in the detector and/or associated parts.

Drift can be caused by improper stabilization of the
electrometer, improper function of the flow control
system of the chromatograph, or elution of high boiling components from the column. Although the FID is
generally insensitive to changes in flow rate, drift
will occur if the change in flow rate brings more or
less ionizable material into the FID. Disconnect
the detector input to the electrometer. If drift
continues, the problems lies in the electrometer/
recorder system and the isolation of the faulty
component may be done as described earlier in this
chapter.

Occassionally it will be observed that attenuation
of the signal will not reduce the signal level by
exactly one-half (for a binary attenuator). This is
not the fault of the electrometer, but is caused by

improper adjustment of recorder zero. If the recorder zero is below the scale zero, attenuation will apparently cut the signal by more than one-half, and conversely if the recorder zero is above scale zero, by less than one-half.

There are two other points which should be mentioned with regard to electrometer operation. The amplifier has a certain limit regarding the output voltage which it can deliver. This will probably be in the range of 10 volts for most amplifiers. If the input signal requires that the amplifier develop more than this limit to drive the feedback current to the grid through the input resistor, the amplifier will saturate. Any signal requiring more than this limit will not be reflected in a greater deflection of the recorder pen. This is indicated by flat-topped peaks at some point less than full scale deflection (usually on some high attenuation setting). This should not be confused with the flat-topped peaks which result when the recorder is driven full scale and is mechanically stopped. If saturation is suspected, switch to a higher attenuation to keep the peak (s) on scale and check for flat-topped peaks. This situation will ordinarily only be encountered on the most sensitive ranges; the higher resistance value requires a higher voltage for a given current. Since the amplifier has a limit of 10 volts, any signal greater than 2×10^{-10} amperes would result in saturation and a square peak at about 75% of full scale. In such a case a lower range position should be used. The second point is that the full potential developed by the amplifier is not dropped across the variable portion of the attenuator. In addition to a fixed resistance of 3,750 ohms in the attenuator, an additional resistance element is used in the attenuator network so that only a few per cent of the amplifier output is dropped across the variable portion of the attenuator. The use of the large fixed resistor in the attenuator output does not affect the sensitivity of the system. The attenuator setting is thus the fraction of this small percentage which is being used to drive the recorder.

Functional Operation of the Thermal Conductivity Detector.

An understanding of Ohm's law is also used to see the

185

origin of the voltage developed by a thermal conductivity detector (TCD). Although an electrometer is not used with the TC detector, a knowledge of the electronics involved should be helpful in diagnosing problems with TCD cells.

A typical thermal conductivity detector-Wheatstone bridge circuit is shown in Figure 12-3. The recorder

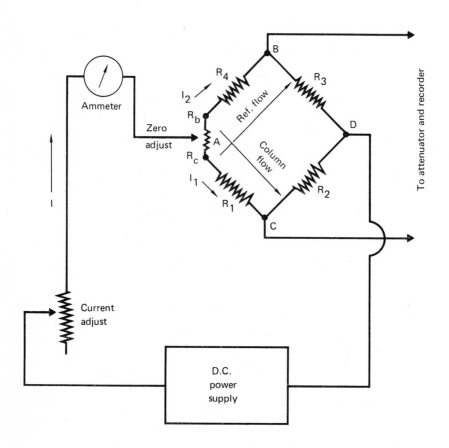

Figure 12-3   Typical Thermal Conductivity Detector Wheatstone Bridge Circuit (9)

is connected so that the potential difference between B and C is being measured. The variable resistance in the "current adjust" section is used to vary the total current (I) flowing through the system. This, in part, determines the sensitivity of the set-up. The "zero adjust" control divides the total current (I) through the upper and lower branches (I_1 and I_2). Due to the conservation of electrical charge (Kirchoff's First Law), $I = I_1$ plus I_2, the zero control is used to vary I_1 and I_2 so that the potential drop AB equals the potential drop AC at the null point, and points B and C are at the same potential (no voltage difference-bridge output = 0). Resistances R_3 and R_4 (reference side of TC cell) are constant: R_1 (column side of TC cell) and R_2 are constant also, but can be variable when a sample passes by R_1. The sum of the resistance in each branch ($R_b + R_3 + R_4$ or $R_c + R_1 + R_2$) is thus variable, but equal. Hence, at the null setting I_1 and I_2 are constant. As a component is eluted, R_1 usually increases. Since the potential at B is constant ($I_2 R_4$ constant) and the potential at C is $A - I_1 R_1$ (due to change in R_1), a voltage difference is developed between B and C. A recorder connected between B and C can then be used to measure the potential difference. The two electrical factors which affect the sensitivity are I_2 (hence I) and the temperature coefficient of resistance of the sensing elements. These are not mutually independent; a higher current would produce a greater response, but a hotter filament would result which would produce a smaller temperature change for a given amount of heat and filament life would be reduced. For problems and solutions, see Chapter 11.

Recorders

The second link in our data system is the potentiometric recorder. It is the function of the recorder to reproduce in a graphic form the output of the electrometer-amplifier or thermal conductivity detector. Figure 12-4 is a block diagram of the components of a continuous balance potentiometric recorder. The 60-cycle line voltage and the chart drive mechanism are independent of the measuring circuit. Since a chromatogram is a presentation of response versus time, the only purpose of the chart drive system is to provide a continuous movement of chart paper in some known relationship of inches of paper per unit time.

187

The differential amplifier compares the input signal from the amplifier-attenuator network with a reference voltage generated in the measuring circuit of the recorder. The difference between the two signals is called the error signal. This error signal is converted from D.C. to A.C. and amplified to a voltage large enough to drive the servo balancing motor. If the difference is positive the motor is driven in one direction and if negative, in the opposite direction. The balance motor is mechanically linked to the reference slidewire and to the recorder pen. The change in the slidewire contact position changes the value of the reference voltage in such a way as to reduce and maintain the error signal at the amplifier at zero. In essence the system continuously attempts to maintain the error signal at zero. The D.C. voltage across the slidewire is the range of the recorder, for example 1, 5, 10, 50 mV. If the input signal is greater than the reference voltage span, the recorder will be driven full scale and will be mechanically stopped.

The accuracy of the measurements of potential in the millivolt range depends on a precise current flow of known magnitude through the calibrated slidewire (Ohm's law). This constant current supply is provided by a constant voltage unit in the recorded (usually a mercury battery). The constant voltage supply sometimes consists of an A.C. to D.C. power supply and a very accurate voltage regulator. Even in the presence of fluctuations in the line voltage the regulated voltage supply will provide the proper current output to maintain the correct milli-voltage drop across the slidewire. Another component of the recorder is the measuring circuit. This section permits the selection of the span of the recorder, electrical zero, and damping adjustments. Discussion of these points will be given in subsequent paragraphs. The amplifier section of the recorder consists of the components necessary to convert the D.C. input into A.C., amplify the A.C. signal, and feed the output to the balancing motor. The chart drive system consists of a constant speed motor and gear network to move the chart paper at a constant rate. Thus, in Figure 12-4, the measuring circuit unit and the constant voltage source would be part of the reference voltage-slidewire combination and the other essential parts are as given in the figure.

188

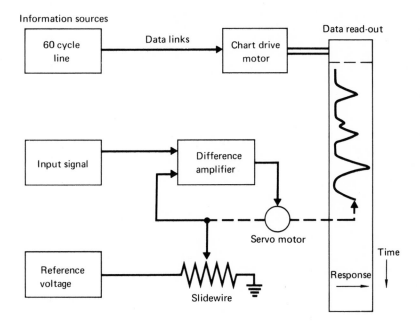

Figure 12-4   Block Diagram of Automatic Potentiometric Recorder (9)

Consideration now is given to some of the terms which
are used to characterize recorders. An understanding
of these is important in maintenance and troubleshooting   of recorders. The terms range and span of a
recorder are related. The algebraic difference between the end-scale values of the recorder and range
is the region covered by the span by specifying both
end-scale values. If we speak of a one millivolt
recorder, we are actually saying that the span of the
recorder is one millivolt, from zero to full scale.
Many recorders will also accept and balance against
a signal below zero- so a more precise statement of
recorder operation would give the range also.

Dead band is defined as the range through which the
measured quantity (the input) can vary without causing response. It is commonly expressed in per cent
of full-scale deflection. Determination of the dead
band of a recorder is a fairly easy operation. A

D.C. signal (usually with a millivolt box) is connected to the input terminals of the recorder so that
the pen is balanced some place off of zero. The pen
is now physically forced a few per cent away from
this balance point by turning the motor pin on gear.
Now release the gear and allow the pen to rebalance.
Then force the pen an.equal amount in the opposite
direction, release, and allow to rebalance. The
difference between the two rebalancing points, expressed as a per cent of full scale, is the dead band.
A dead band in excess of 0.5% of full-scale will
cause faulty recorder operation, especially missing
of very small peaks in capillary GC operation.

The input impedance is the sum of the resistance and
impedance of the input circuit of the recorder; it
is of little value to the user. What is more important is the source impedance. Source impedance
is the resistance the recorder "looks at" in the
attenuator output, and this must be in the range
specified for a given recorder for linear operation.

Pen step response speed is defined as the time it
takes for the pen to come to rest in a new position
after an abrupt change to a new constant value in the
input signal. If the recorder is balanced at zero
and suddenly a one millivolt signal is imposed on the
recorder, the time between the application of the
signal and the time the pen comes to rest at one
millivolt (e.g. full scale for a one-millivolt recorder) is the response time or pen speed of the recorder. A value of one second is common. Slewing
and damping are the terms used to specify the deceleration of the pen. Since the chart pen represents
a dynamic system in search of new rest point, overshoot of the rest point will occur unless some means
is provided to decelerate the pen as the balance
point is approached. If the pen is over-damped, the
pen will only gradually approach the final balance
point. Many recorders have an RC-circuit in the
system which provides a means to anticipate the final
balance point and cause deceleration before this
point. The damping adjustment controls the variable
resistance in this circuit.

Interference rejection is a measure of the recorder's
ability to operate without change of dead-band or
calibration in the presence of extraneous A.C. sig-

nals. An alternating current has the same effect on the input terminals of the recorder as it would on the input signal to the electrometer, that is it would create a current flow because of the changing magnetic field associated with the A.C. field (generator effect). The main source of A.C. interference is from the A.C. power lines into the instrument and hence the stray signal will have the same frequency as the power line (60 c/s). Most commercially available instruments have filtering circuits and shielded inputs to eliminate or reject extraneous signals. These provide a barrier to the higher frequency stray signal and paths through which the signal can flow to ground rather than to the recorder.

The gain or sensitivity adjustment of the amplifier is needed to reduce the dead band of the recorder. Since the measured signal is in the millivolt range, the error signal (input signal minus reference signal) will be in the micro-volt range and a large amplification of the error signal is necessary. To adjust the gain, turn the gain adjustment control until the pen starts to oscillate, then reduce the sensitivity slightly (turn the control in the reverse direction).

The damping adjustment is a means of varying a resistance in a RC-circuit as previously mentioned. The damping adjustment normally is included in the measuring circuit panel. To adjust the damping requires application of a fractional milli-volt signal. The response of the pen is observed upon application of a 10-50% span signal. If the pen overshoots, the system is underdamped or if the pen undershoots (is sluggish), the system is overdamped. Trial and error is needed to bring the pen to the final rest point with the least over-or under-shoot in the minimum time.

Slide-wire cleaning should be a routine, perhaps weekly item. Naphtha, or other solvents leaving no residue (such as methanol) can be used. A stiff brush should occasionally be used to remove deposits which accumulate between the convolutions of the slidewire. At this time inspect the slidewire for signs of excessive wear, especially in one spot. Since there are many different types of writing mechanisms, pen cleaning may take several forms. It may only require replacement of a ball point pen or replacement of a solid ink supply used in some recorders. Best bet on this point is to consult the recorder manual.

Recorder Problems and Solutions

No response can mean that the recorder is not plugged in (simplest case), the power is not turned on, the input leads are not properly connected, or, in some cases, the chart drive is turned off, as in some instruments the chart drive switch also inactivates the measuring circuit. Mechanical binding in the balance motor or the pen carriage can also prevent response. Check the servo motor to make sure it is free to turn. No response or sluggish response may be due to a faulty amplifier. Check the amplifier tubes.

If the instrument is driving against a stop full-scale, check the output of the constant voltage circuit, the measuring circuit, or the amplifier according to the manual. This may also be an indication of improper shielding which is allowing a stray signal to be impressed on the input terminals. Check for continuity of shielding. Also check the incoming signal from the electrometer or RD cell to see whether some large voltage output may be overdriving the recorder.

Sluggish response may be due to the gain adjustment set too low or overdamping. Check these points as listed above.

Erratic or jerky pen movement is an indication of a dirty or uneven slide-wire, or a worn contact, mechanical binding, or under damping.

Non-linear attenuation of signal was discussed previously. Check the recorder zero in the absence of an input signal. Turn the attenuator to infinity or remove input leads.

Pen oscillation indicates gain too high, or damping too low. A poor ground connection or a loose input connection will also cause pen chatter, as will a worn slidewire. Check to see if the oscillation or chatter is dependent on the position of the pen along the scale. If so, check for a worn spot on the slidewire at the position of noise.

Failure of the pen to return to baseline after a peak is due to a large dead band if associated with a recorder problem.

Erratic chart drive is due to slippage in the chart drive clutch or slippage of the gears.

If the pen searches all over the chart without finding recorder zero or over-responds to chromatographic signals, this usually is associated with failure of the battery in the recorder reference circuit, and the battery (usually a mercury constant-voltage cell) must be replaced.

The recorder manual provided with the instrument will normally contain a section describing the common symptoms and remedies. Thus, the last point to be made should probably have been the first - READ THE MANUAL. A recorder checker is described in Appendix E.

Integration Techniques

Quantitative analysis of the chromatogram requires that some quantity which varies with the amount of component being detected be measured. The simplest quantity to measure is the peak height since this can be read directly from the chart paper divisions or measured with a scale. In many applications peak height measurements are sufficient. The technique does, however, require precise calibration of the instrument by injection of varied amounts of the component (s) of interest under identical conditions. If the analytical conditions are maintained constant, the peak height will be a direct measure of the component within the linearity of the detector. A quantity more amenable to automatic measurement and better quantitation is the peak area. The triangulation method (H x W 1/2) is time consuming and is subject to human error. The more measurements that have to be made, the greater is the chance to make a mistake.

The technique of cutting out and weighing each peak suffers from the same disadvantages; time consuming, subject to error, dependence on accuracy of cutting, homogeneity of paper, accuracy of weighing, etc. All that is required is a pair of scissors a lot of patience, and an $800 analytical balance. For this price or less you can buy a disc integrator.

Similar arguments can be used against the planimeter. From an initial investment standpoint, it is cheap, but from an operational standpoint it is quite expensive since it is time consuming and frought with sources of error (mainly human).

The operation of the disc integrator has been fully described in the literature (4) and on manufacturers' brochures. One of the features of the disc integrator which is highly objectionable is that it is dependent on the recorder operation and any trouble in the recorder will also affect the accuracy of the integration. Change of baseline is not automatically compensated and a correction between peaks (zero count rate). Each peak must be attenuated to a peak height less than full scale, which may require a trial run on each sample. For the analysis of fairly simple samples with stable analytical conditions (no baseline drift or shift) and a predictable composition, and disc integrator is a good investment. Little can actually go wrong with a disc-type integrator as long as the mechanical connections to the recorder measuring pen are intact and functional. This should be checked daily.

The Daystrom integrator was marketed in this country, but is no longer available. It is included only to point out the principle of integration. Integration was based on the charge (Q) developed on a capacitor with applied voltage (V) ($Q = C \times V$, where C is the capacitance) and was somewhat automatic. The signal was automatically attenuated and the capacitor discharged when the peak reached full scale. The capacitor discharge was indicated by a pipping pen and the residual charge on the capacitor at the termination of a peak was recorded as a deflection of the recorder pen as some per cent of full scale. The area of the peak was the sum of the pip plus the size of the final deflection. Thus, as with previous methods, line counting was required to translate the integrator output into a number proportional to peak area. No maintenance or troubleshooting hints will be discussed for this type of integrator.

Digital Integrators now provide completely unattended integration of the chromatogram. The aim is to provide the analyst with a digital read-out which is a measure of the peak area. The only errors will be in the ability of the analyst to relate this number to the composition of the sample and the fidelity of the detector-electrometer-integrator combination to produce a number which is truly indicative of the concentration. The integrator will produce a number which varies as the size of the electrical input that it sees. The fact that the integrator gives a printed number does not give gas chromatography any more accuracy that the analyst is able to build into his analytical scheme. The fact that this number may be larger than the output of other integrators likewise does not improve the accuracy of the technique.

For complete automatic operation the integrator must have certain features. There are several instruments on the market today which include all or part of these features. One required feature is a high count rate, expressed in counts per second per millivolt of input. Coupled with the count rate must be the ability to print out a large number of counts, since a high count rate may produce an output in the tens or hundreds of thousands of counts for large peaks. A high count rate is also essential so that good statistical reproducibility can be obtained on small peaks. The linear dynamic range of the integrator is the ratio of the largest to the smallest peak which the instrument can handle within the system's linear region of response. If the integrator can give a linear response to a 50 mV signal as well as a 10 microvolt signal (0.01 mV), the linear dynamic range (L.D.R.) is 50/0.01 or 5,000. Extension of the L.D.R. is accomplished by auto-ranging: as the signal approaches a certain preset level, the instrument automatically reduces the amplifier sensitivity down by a factor of ten and also automatically feeds the output signal to the next higher decade range of the counter.

The ability of the integrator to handle a variety of input signals encompasses many variants, some of which are shown in Figure 12-5.

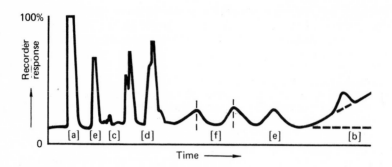

Figure 12-5   Various Chromatogram Situations Encountered by Digital Integrators (10)

The input signals may range from a few microvolts to
a few hundred millivolts. a) Baseline instability
caused by drift must be recognized as such and compensated for by the integrator. b) Noise, usually
higher-frequency noise, must be rejected by the
integrator. c) Incomplete resolution of components
is a common occurrence in GC and presents another
challenge to the integrator. d) The incompleteness
of resolution may manifest itself as peaks of the
same or vastly different sizes or peaks with or without a defined valley between them (shoulders).

The first peaks eluting from the column are normally
narrow and quite sharp, whereas later peaks may rise
rather gradually and be many seconds or even minutes wide, depending on the analytical conditions.
e) With capillary columns the earlier peaks may be
only a few seconds wide, while the last peak may be
30-60 seconds or greater in width. In addition to
a varied peak width certain peaks may have a definite
asymmetry, with either a sloping front or back f).

Superimposed on the D.C. detector signal will be some
A.C. signal from various sources. The integrator
must have some means whereby we can control the response to these A.C. signals or a filtering network
to remove them.

Basically the logic of an automatic integrator uses
three parameters for detection of peak on-set: the
slope of the signal (rate of change of millivoltage
per second), the amplitude or size of the signal,

and the frequency of the signal. If the signal is
recognized as the on-set of a peak (see Figure 12-6),
the amplified signal is sent to a voltage-to-frequency
converter whose frequency output is directly proportional to the output of the amplifier. The frequency
pulses are set to the appropriate decade of the data
counter and a visual display of the accumulation
may be provided. The peak sensor logic also includes
provision for signalling the end of a peak at which
time the accumulated count in the data counter is
printed out in digital form. For identification
purposes circuitry may also be provided to detect
peak maximum (slope goes from positive to zero to
negative) and store in a memory system the time which
is printed out along with the data count. The read-
out consists of a retention time count (up to 999
seconds from start) plus 5-8 digits of area data.

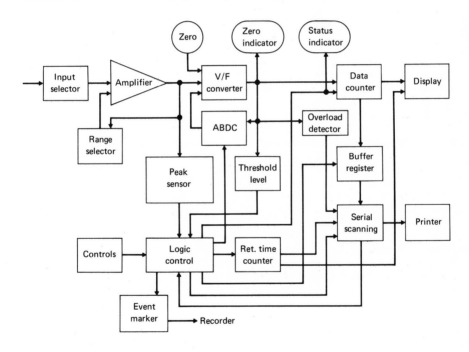

Figure 12-6 Block Daigram of Automatic Digital
             Integrator (Courtesy Infotronics Corp.)

Slope detection works on the change of input signal from zero to positive (increasing signal) to detect on-set and change from negative to zero slope for peak end. This eliminates an absolute reference level and allows integration of overlapping peaks. Provision may also be provided for integration when the slope goes positive to zero to positive(front shoulder) or negative to zero to negative(back-side shoulder). Frequency filtering is used to eliminate high frequency noise or spikes. Amplitude or trip level adjusts the level to which the signal must rise before being recognized as a peak. The value of the level control is the percent of full scale to which the signal must rise before integration commences. For low noise level systems a low value of 1-3 is used. This insures that even small signals will be integrated and increase slope sensitivity.

Correction for baseline drift may also be included. This device works in conjunction with the slope and amplitude controls. If drift above the trip level occurs at a slope less than that required for the logic to recognize it as a peak, the drift correction automatically raises the reference level or baseline. It does not affect the output to the recorder so the drift still appears on the chromatogram. This feature is inoperative during the elution of a peak and hence a certain error will exist if the drift is severe and the peak is fairly broad. At the end of the peak the drift corrector will again commence operation and the base level will be brought to that existing at that time. This feature is of no value for closely spaced peaks on a continual drift since it will not work except during the time between peaks (minimum of 2-5 seconds needed to sense baseline).

There are also options available known as exponential decay correctors for accurate integration of peaks occurring on a severely tailing peak, for example from solvent.

Characteristics of Commercial Integrators

A typical digital integrator has a solid state v/f converter input system which has slope, amplitude and frequency peak sensor logic. A slope as small as 0.01 uv/second can be sensed as a peak. The majority have a linear dynamic range of 500,000, which is centainly adequate for most situations.

Typical count rates of about 2,000 c/s/mv with a
maximum count-rate of about 10^6 c/s with a capacity
for 8 digits is common (00,000,000). A three digit
retention-time readout is standard. Most units
can handle two peaks/second if necessary. An auto-
ranging device allows inputs up to 500 mV, with
auto-ranging occuring at 50 mV. Such integrators
can easily handle all commonly encountered GC input
signals. Many output options are available including paper tape, magnetic tape, card-punch, and computer.

Today's digital integrators are varied and very
complicated pieces of electrical equipment, and actual troubleshooting by the laboratory analyst is
difficult other than the correct setup of function
controls as described in the user's manual. However,
certain routine maintenance procedures can be followed to keep the integrator operating efficiently.
Unless the integrator is not going to be used for
long periods of time(longer than one month), do not
turn the electronics off, leave the unit in the reset
position. This helps keep the internal electrical
components at a constant temperature, and thus, helps
prevent failure of components due to temperature
changes. The ventilating-fan area should be kept
clear of accumulated dust and dirt, as this can
cause undue temperature buildup in the cabinet, and
result in thermal failure of solid-state and/or tube-
type components. The only integrator failure one
of the authors has had with a digital integrator in
six years of operation was due to a buildup of dirt,
and poor air circulation within the cabinet, which
resulted in a few dirty contacts that were rapidly
cleaned and resoldered. Also, be sure the integrator
is placed in an area that allows good air circulation.
Removal of the top of the cabinet or sliding out the
equipment chassis (depending on design) will allow
dirt and dust removal from the interior of the unit
by directing a fine stream of air on the dirty areas
and prevent malfunctions in these areas due to dust
accumulation. Other repairs, except for checking
and changing fuses and function lights when required,
should best be left to the experienced service engineer.

Computer Linkups

In this section only the equipment available will be
discussed, as troubleshooting these systems are for
highly trained technicians, not the average chromatographer (including the authors). Very few analytical laboratories have the facility for on-line
computer handling of GC analytical data due to the
huge capital expenditure required to purchase the
computer components. Only the very largest of laboratories and modern manufacturing plants have such
a capability. Also, there are many laboratories
in which the routine GC analysis load is not sufficient to justify even a minimal on-line system.
However, there are workers involved in non-routine
chromatographic research who could benefit greatly
from computer automation, but for a variety of
reasons (mainly financial) cannot go on-line with a
computer having sufficient computing power and/or
data-storage capacity to meet their particular needs.
In such situations as these, off-line data aquisition and control systems can be utilized to permit
the gas chromatographer to realize the benefits of
computer data handling at a cost he can afford. As
a matter of definition, the on-line computer is interfaced directly to the GC unit, while the off-line
system produces a recorded digital output which is
submitted for computer processing at a later time(5).
The digital output data to be stored is usually produced via magnetic or paper-punched tape, or a card-
punch. In this section we will briefly discuss off-
line systems, on-line time sharing systems, small
dedicated computer systems, and in-house on-line
computer system.

The primary feature which characterizes the off-line
computer system is the indirect link between the
digital output of the unit and the computer. A second feature is the lack of a direct control like from
the computer to the GC system and the presence of
a system control module. Off-line systems are further
characterized by modularity. The chromatographer can
choose from a variety of data aquisitions systems
as well as from numerous computer facilities. The
system usually consists of a dedicated coupler, i.e.,
specially built interfaces to measuring instruments
and recording devices such as paper-tape punch,
magnetic tape, or IBM card punch. After the data

are recorded, the tape or cards are taken to the computer for processing and printout. One of the main
advantages of this method of data aquisition is that
the tape and/or cards provide a permanent record of
the data, as well as being quite simple to operate.
The recorded data may be then taken to a large or
dedicated computer for processing, or if paper-tape
is the data gathering medium, to a time-shared computer terminal.

For routine analyses, the analog to digital conveter
most widely applicable to off-line chromatographic
systems is the digital integrator. As discussed
previously in this chapter, such units are available
commercially with interface circuitry installed to
allow direct installation with various off-line data
aquisition systems as well as on-line with computers.
If a time-sharing computer is available in the laboratory, a common teletype facility with keyboard
printout and simultaneous paper-tape punch can be
used. The retention time and peak area data are
printed out, as well as being punched on paper tape.
The data on the paper tape is then fed in to the time-
share computer. The only drawback with this system
is that the teletype terminals yield slow data-print-
out (10 characters/ second), and for some rather
involved programs for calculating solution properties
from compositional data, it would take as long as
two-three hours to print out the data. This could
be expensive as well as nervewracking. However, for
most routine-type analyses, this is an ideal off-line
system. Another often used system for data aquisition
is magnetic tape. This is a particularly good system
for routine analyses from which detailed calculations
are required, as most large computers have a magnetic
tape data input system. The last system for off-line
data aquisition to be discussed is the IBM card
punch. There are easily interfaced with commercial
digital integrators and result in a data card deck
for each analysis, with an individual data card for
each peak in the chromatogram. Such a system is invaluable in the analysis of a complicated mixture
such as motor gasoline, as each card (representing
a peak) may be identified by a code number to obtain
the detailed composition of a gasoline as described
by Sanders and Maynard (6). To process a data card
deck, a large computer is usually required, and
such a system is only really useful when detailed

calculations are required that need a great number
of characters printed-out. The other previously
mentioned data aquisition systems are more practicable
for routine analyses.

With the advent of time-sharing computer access being available in almost any area where long-distance
telephone service is available, the capability for
on-line usage of a remote computer is possible. Dial-
ing the computer's phone number puts the user in
direct communication with the computer, and he is
charged on the basis of computer time used (usually
$5 - $20 /hour + fees). The program data are permanently stored in the computer and are recalled in
a fraction of a second to do the desired calculations.
The interface between the computer and the digital
integrator is usually the teletype. The advantages
of such a system is conversational data reduction
of on-line recording, permanent record and printout
of results, and simple to operate. Also, the softwave required (programs) for most calculations are
quite simple, either written in FORTRAN or BASIC
(a conversational language, easily learned by some-
one with no previous computer experience). The major
drawback is that only one chromatograph can be connected to a teletype terminal at one time (although
several can be connected simultaneously to the remote computer with separate teletype units) and the
type of analyses handled would have to generally be
routine that the computer has been previously programmed to handle, with peak identification by elution
time or by a retention time index techniques. Such
a system would not be very useful for quantitative
analyses of non-routine samples, but would be quite
useful in non-routine analyses where a sophisticated
computer is required for complicated chromatographic
interpretations (for example, the evaluation of phy-
siochemical parameters from measurements of central
moments of the chromatographic peak profile (7).

Dedicated computer systems generally use small computers (although IBM 1800 systems are common) that
process data in "real time" (simultaneous with the
analysis) providing information immediately for the
researcher in the laboratory. Such units are dedicated to process only the type of data they are programmed for, (hence the name), and must be reprogram-
med for any different type of analysis (as does any

202

computer). However, due to their small capacity, dedicated computers can only handle a limited number of analyses. The main advantages to these types of computers are high recording speeds (to 10kHz characters/second), immediate display of results, and on site control of the GC unit parameters, such as column oven control, sample injection, etc. The main disadvantages are: 1. programming skills are required to operate effectively, 2. generally no permanent input data record is available and, 3. the systems can be very complex and expensive. Thus, unless built especially for handling GC data, (and there are several such units now commercially available), the small dedicated computer would be difficult for the average chromatographer to handle. However, units built specifically for monitoring GC digital integrator outputs, and not for controlling analysis parameters, are easy to operate and especially attractive for the GC expert who can spare 10K - 50K dollars for the unit.

The in-house, on-line large computer system is that which is uselessly dreamed of by many chromatographers, especially if their lab is small, because of the prohibitive cost of large computers. Although, such computer linkups are of great value, they are not a panacea to the chromatographers data handline woes. Even though instant results are achieved at the end of an analysis, and the computer can be programmed to control the GC analysis parameters, the analytical data are not recorded, and often an analysis must be repeated only because the results "appear" to be in error. This type of linkup should probably produce better quantitative results than the digital integrator, as a much better job can be done in correcting peak areas of partially resolved peaks and proportionally correcting for a drifting baseline. However, it requires a good knowledge of programming to even begin in this area (or immediate access to a Systems Analyst), and still the computer could only positively identify peaks with regard to peak patterns, elution times, and retention indicies. Such softwave techniques can be used for isothermal and programmed-temperature runs of relatively simple mixtures; however, for extremely complex wide-boiling mixtures (such as gasolines) that require broad temperature programs which are difficult to repeat exactly, the computer can be fooled, and peaks misidentified. Regardless, at some time in the future, the on-line

computer will probably replace completely the off-
line systems, and it is important that all of us
with responsibility in GC analyses remain aware of
developments in on-line techniques. However, in the
near future, off-line data aquisition remains the
most attractive means of automatic data handling for
the majority of laboratories. While on-line systems
have the potential ability to produce better analytical data than the digital integrator, this has not
been conclusively borne out by experimental evidence
(8), and the experience we gain in using off-line
systems should be quite valuable when the great tran-
sition occurs.

In this chapter we have discussed the function and
operation of various components in the electrometer/
recorder/digital integrator (or GC data producing)
system, and described some routine maintenance and
troubleshooting techniques. The interfacing of GC
data producing systems with computers was also discussed, but due to the complexity of these interfacial
arrangements, no maintenance or troubleshooting
procedures were discussed, as this could be the subject of another entire book, not just a small part
of one chapter. The analyst can often troubleshoot
and repair electrometers and recorders (with careful
reference to the users manual); however, troubleshooting of complex solid-state digital integrators
should be left to experienced service engineers.

Chapter 12

References

1. L.S. Ettre, Open Tubular Columns in Gas Chromatography , 1st. Ed., Plenum Press, New York, 1965 (p 133).

2. Hewlett-Packard Detector and Controller Manual, 1970, Hewlett-Packard Corporation, Avondale, Pennsylvania.

3. Varian-Aerograph Model 1520 Users Manual, 1969, Varian-Aerograph Corporation, Walnut Creek, California.

4. S. DalNogare and R.S. Juvet, Gas-Liquid Chromatography-Theory and Practice , 3rd. Ed., Interscience Publishers, New York, 1965, p.255.

5. J.G. Peddie, "Research and Development", April, 1971 (p. 26).

6. W.N. Sanders and J.B. Maynard, Anal.Chem.,40, 527 (1968).

7. O. Grubner, Advances in Chromatography , Vol. 6, New York (1968).

8. J.E. Oberholtzer, Journal of Chromatographic Science , Vol. 7, December ,969 (p. 720).

9. J.B. Maynard, Private Communication, 1971.

10. M.T. Jackson, Private Communication, 1971.

Chapter 13

Auxillary Systems

Since the beginning of separation of organic compounds by GC, there has been a continuing search for simple qualitative techniques to identify the minute amounts of components separated by GC other than their emergence time from the column. Many methods have been employed to accomplish this difficult identification task. In this chapter we will discuss several of the more successful auxillary systems used for qualitative monitoring of GC columns as well as some of the problems that are often encountered in their use.

In the late 1950's, a popular method to characterize the components emerging from the GC column was by functional-group wet-chemical analyses. In this technique, the column effluent (usually after passage through a thermal conductivity detector-TCD) is bubbled through a small amount of the desired reagent to determine if a given functional group is present in the peak under scrutiny. For example, an acidic 2,4-dinitrophenyl-hydrazine solution is used to determine whether a peak contains an aldehyde or ketone carbonyl compound. If such a carbonyl is present, the usual 2,4-dinitrophenyl hydrozone derivative will be formed as a precipitate in the solution. The precipitate can then be dried, recrystallized (if enough material is present), and a micromelting point determined to help in identification of the peak, as the melting points of 2,4-DNPH derivatives are characteristic for many of the various carbonyls. Another characterizing reagent often used is bromine solution in carbon tetrachloride, which discolors from the reddish-brown of bromine to a clear solution in the presence of olefinic double bonds. It has been the authors experience, that at least 20-30 μg of component be eluted in a given peak to get a positive reaction with most functional-group reagents, although the limit of detection quoted for some of the functional group tests is much lower. To obtain useful results with these systems, 1/4 inch o.d.

packed columns are usually employed to allow large enough sample sizes. A detailed description of a useable GC sampling technique, reagents and examples has been reported previously by Walsh and Merritt (1).

The major problems of concern with this type of system is the same as with any other micro wet-chemical system. All gas-bubbling vessels should be kept clean to prevent interference with the primary reaction to be used. Usually a small bore glass or stainless steel extension is attached to the TCD outlet and extended to the bottom of the glass reaction vessel. This tube must be kept very clean, and warm, if possible, to prevent any condensation of heavy components emerging from the chromatograph. The tube should be removed and periodically cleaned with a polar solvent (such as chloroform) to ensure no cross contamination between samples examined. If the above mentioned cleanliness suggestions are followed, very little can go wrong with this simple technique for monitering column effluents. One suggestion is that the very slowest flow rates possible should be used to allow the reagent the maximum time to mix and react with the materials in the column effluent.

The most successful auxillary techniques for qualitative identification of GC peaks have been spectroscopic in nature including infrared (IR), ultraviolet (UV), and mass spectroscopic (MS) techniques. Of the three methods mentioned, the GC/MS technique has been highly refined from the beginning with a Bendix Time-of-Flight mass spectrometer (TOF-MS) monitoring the effluent of a packed column (2) to present day monitoring of capillary column effluents with fast-scanning, high-resolution mass spectrometers. In addition, the methods for interfacing GC/MS systems have been greatly improved, and have been reviewed in some detail by Rees (3).

Infrared Spectroscopy

Analysis of GC fractions by IR gives considerable functional group and structural information on the fractions. One of the better methods for obtaining IR spectra of GC peaks of components with low vapor pressure is to trap the peak in a cooled (dry ice or liquid nitrogen) flow-through trap, transfer the contents of the trap to an IR cell, and then record the IR spectrum in the usual manner. Another method is   to trap GC fractions eluting through a nondestructive detector(such as the TCD) using an ordinary, straight melting-point tube, with about the middle 2-inches of the tube being cooled to dry ice temperature. The same type of trapping can be done from a column split-stream if a destructive detector such as an FID is used. The melting-point tube is usually connected to the GC effluent with a specially machined piece of Teflon tubing that gives a leak-tight seal. After the component has been trapped, the exhaust end of the tube is capped with a silicone rubber septum, the tube disconnected from the Teflon connector, and that end quickly capped with a silicone rubber septum to prevent loss of inert carrier gas and the possible entrance of air and water vapor into the tube. The melting-point tube is then removed from the dry ice and allowed to warm to room temperature. A septum is then removed from one end of the tube, and the open end inserted into a micro-IR cell(NaCl) as shown in Fig. 13-1. The sample is then washed from the sides of the tube with a small(1-5 μl) amount of chloroform or carbon tetrachloride into the micro-IR cell sample cavity. If enough sample is present in the trapping tube, the entire assembly may be placed in a small laboratory test tube centrifuge(cushioned in the bottom by tissue or cotton to prevent fracture of the microcell during centrifuging), and the liquid in the tube centrifuged into the IR-microcell sample cavity.

When the sample is in the IR-microcell, a small drop of mercury is placed in the cell above the sample to prevent loss by evaporation and to keep moisture from entering the cell. The surface tension of the Hg is such that it will not drop into the sample reservoir. The cell is then placed in a special microcell holder, and with the aid of a beam condenser in the reference beam, the IR spectrum obtained in the usual way. Commercially available are devices for   trapping peaks in KBr, and obtaining their spectra in a KBr pellet; however, such treatment is only useful with very heavy components.

The major pitfalls of using the microcell technique

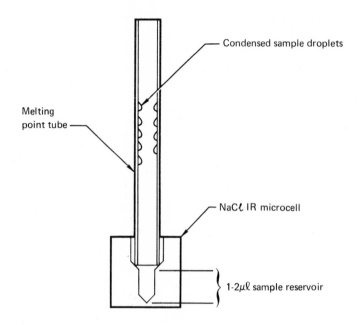

Figure 13-1   Sample Transfer from Melting Goint Tube
              to IR Microcell Diagram

previously described or any other sample cold-trapping
method for IR is loss of sample due to inefficient
tubes used for trapping. (this is especially true if
the sample would tend to form an aerosol in the
carrier gas at dry ice temperature). Another problem
is that of contamination with closely eluting components or even something absorbed from the atmosphere. It is of utmost importance that the cell
and trapping tube are washed with reagent grade solvents (chloroform usually used) and dried before
use. Also, it is important to cap the tube tightly
under the slightly positive carrier gas pressure
available at the column exit to minimize the entrance
of water and CO_2. Water will not only interfere
with the IR spectrum, but will damage the NaCl
microcell. CO_2 will interfere with the IR spectrum.
To prevent this, the caps should not be removed from
the trapping tube until it has essentially reached
room temperature. The sample transfer procedure

should then be carried out as rapidly as possible.
If the microcell technique is used, it is also important that there are no leaks around the Teflon
sleeve used to connect the trapping tube to the GC
unit outlet. A loose fit will allow back diffusion
of air and moisture into the cold trap due to its
reduced temperature.

For more volatile components, samples can be taken
in evacuated, longpath gass cells. As the peaks
are eluted, the column effluent is diverted to the
gas cell, and at the termination of the peak the
cell is removed and sealed-off by closing the inlet
and exit valves. Most gas cells of this type can be
heated to about 200°C, if necessary, before recording the IR spectrum. The spectra obtained in this
manner (gas phase) shown more fine structure because
the molecules have more freedom of movement in the
carrier gas matrix than in the liquid phase. A
typical gas cell is shown in Figure 13-2.

The major problems associated with collecting gas samples
for IR analyses occur with loss by leakage, condensation

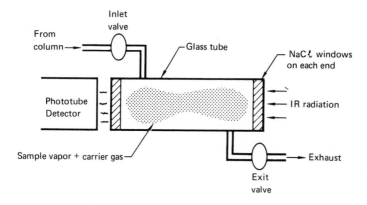

Figure 13-2 Diagrom of a Typical Gas Phase IR Cell

on cell walls, interfering gases (such as back diffusion of water, CO_2 etc. from the atmosphere) and
contamination of the peak being examined by
closely eluting components. Fairly large sample
sizes are generally required to obtain a sufficiently
concentrated gas sample, and as a result, resolution
often suffers.

There are commercial IR units available that are
constructed for direct interfacing with a GC. These
units work on the gas phase principle, can rapidly
scan the region from 1.5 to 15 microns in 30 to
45 seconds, and be ready for another component. The
heated cells (250°C) are constantly flushed with
carrier gas, then when a component appears whose
spectrum is required, the column flow is diverted
to the sample cell with the exit valve closed.
Immediately on capture of the component the inlet
valve is closed, and the IR spectrum recorded.
Between successive samples the IR absorption sample
cell is flushed with carrier gas by opening both
the entrance and exit valves. Infrared spectra of
known compounds are obtained in the same manner as
the unknown for comparison. Two units of this type
available are Beckman's IR-102-GC/IR detector/
spectrometer and the Wilks chromatograph-IR
spectrograph. If sample components are well enough
separated (usually by packed columns or large-
bore capillaries), these units, along with retention
time data, usually give enough information to
identify the component peak, if pure (i.e.-not more
than one component in the peak).

To assure proper operation of a gas-phase IR
system, all connections must be routinely checked
with soap solution to find any leaks that could
result in sample loss or atmosphere diffusion into
the cells. Routine maintenance of the cells is
accomplished by a high-temperature bake-out at the
maximum cell temperature with a carrier gas flow
(or purge gas, bypassing the column) of 100-200 ml/
min. This is sufficient to remove most contaminants
adsorbed on cell walls. If cell construction permits, a further step in cell cleaning is that of
evacuating the gas cell while heating. This approach
is especially useful for cells of stainless
steel construction.

212

Ultraviolet Spectroscopy

Gas samples can be taken for UV analysis in the same
manner as just described for the IR gas phase units.
However, liquid samples are usually diluted in micro-
UV cells with spectral grade normal-heptane or cyclo-
hexane for UV analyses. Such UV spectral scans are
particularly useful if aromatic compounds are
present. The major problems with collecting GC peaks
for UV analysis are sample loss by handling or vapor-
ization (same as for IR samples), and contamination
from outside materials. Also, the presence of
exygen from the atmosphere can greatly mask and alter
UV spectra, especially in the far UV region. Thus,
special precautions must be taken to keep oxygen
out of both the gas phase systems and the liquid
pahse system as this promotes oxidation of the
material to be analyzed. If the presence of exidants
are expected, scan the desired spectral area once,
pause for a minute or two, then rescan. If the spectra
are changing oxidants probably are present and the
spectra will not be representative of the pure compound. Any oxidation of a molecule is significant
in the UV, as the C=O or C-OH groups and other O-
containing groups are chromophoric and tend to mask
information about the hydrocarbon structure of the
molecule.

Mass Spectrometry

Gas and liquid samples are collected in cold traps
for MS analysis as described previously for IR and
UV analyses, except in MS analyses, the sample must
be revaporized, and transferred to the ionization
region of the MS without contamination or loss.
Such systems are necessary if only slow-scanning
MS units are available. Many complaints are registered
by the mass spectrometry laboratory that the gas
chromatographer has just submitted another trap
filled with CO_2 and water, and precautions
should be taken to circumvent this difficulty.

213

A trap filled unexpected contaminants such as CO_2 and water vapor when taken to the mass spectrometry laboratory either indicates a serious leak in the trapping system, or a leak during vapor transfer to the MS. All traps should be pressure tested to one atmosphere, and all connections checked for leaks with soap solution. Traps should be maintained after every use by a vacuum bakeout at 250°C (especially if stainless steel traps are used). One quick way to vacuum test a trap is to flush at high temperature with a dried stream of prepurified nitrogen. Connect this trap to the MS inlet system used, and begin to pump away the nitrogen from the trap. While pumping out the trap, continually take spectra, looking for increases in the concentration of water and CO_2. If no increases in the background level of these atmospheric constituents occurs during trap pump-down, it is vacuum tight, and may be used for trapping from GC columns without worry of atmospheric contamination by trap leakage. However, much better and quicker techniques are available to the gas chromatographer who needs MS information for GC peaks to aid in their identification.

The successful interfacing of GC units with fast scanning mass spectrometers has been one of the most important developments in the qualitative identification of GC peaks. Since sample sizes required for good mass spectra are so small (good quality spectra can be obtained from as little as 10^{-9} gms of sample), the effluent of a capillary column can be split 50/50 between an FID and MS and still obtain a good chromatogram along with mass spectral data for peaks of interest. The beginning of tandem GC/MS work was begun about 12-14 years ago when Gohlke published his work at Dow Chemical Company concerned with monitoring a GC column effluent with a Bendix TOF-MS (2). The Bendix TOF unit has been extremely useful with GC units, as it produces 10,000 or 20,000 separate mass spectra/second in the 0 – 220 unit mass range, depending on the desired frequency of operation. Displayed on an oscilloscope, this appears as the complete spectrum for the eluting component, and can be photographed with a Polaroid-type camera for spectral interpretation. Later units were then fitted with gated-type scanning units, that allowed scanning a decade (mass 20-200) in about 3 seconds, fast enough

214

to obtain a useable spectrum from rapidly eluting capillary peaks. Even at the low resolution available with the old TOF units, the mass spectra, along with retention time data, provided a powerful tool for identification of compounds purified by the GC separation. However, several modern techniques for interfacing GC with MS units now allows GC units to be connected directly to fast-scanning high-resolution mass spectrometers that have unit resolution val es from m/e of 1 to 10,000.(3)

A GC unit carrier gas at the detector exit normally exits at atmospheric pressure with flow rate from 1-200 ml/min. However, most MS units operate best at pressures of 10^{-4} torr or less. Thus, any method to selectivity remove the carrier gas from the sample prior to entrance into the ionization region of the MS is certainly desirable to allow a more concentrated sample to reach the ionization region. Several continuous methods for effecting this carrier gas removal have been used. These "molecular separators" fit into the GC/MS system as shown in Figure 13-3. (4)

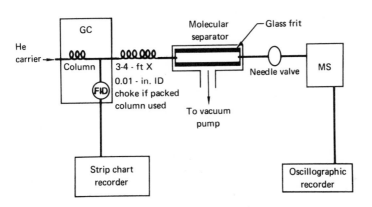

Figure 13-3   Block Diagram of Typical GC/MS Interfacing

Some typical terms used to describe the performance of molecular separators are the Yield (Y) and the Enrichment (N). Grayson and Wolf (4) have defined these separator operational parameters as follows:

1. $Y \text{ (yield)} = \dfrac{\text{sample entering spectrometer}}{\text{sample entering separator}} \times 100 = \%$ value

Thus, the % yield represents the ability of the separator to pass sample molecules from the GC carrier gas to the ion source of the MS unit.

2. $N \text{ (enrichment)} = \dfrac{\text{sample/Helium ratio on MS side of separator}}{\text{sample/Helium ratio on GC side of separator}}$

3. $N = (Y) \left(\dfrac{\text{Helium flow into separator}}{\text{Helium flow into spectrometer}} \right)$

The enrichment (N) may be a large and meaningless parameter if the amount of sample reaching the MS is too small for a good spectrum.

In the following paragraphs several different separator systems will be discussed, along with typical maintenance procedures as well as specific problems and solutions when available.

Fritted Glass (or Biemann-Watson) Separator

This is an all-glass separator introduced by Biemann and Watson,(5) (6) and consists essentially of a porous glass tube which acts as a stream splitter, with the faster diffusion of the small carrier gas molecules through the fritted (porous) glass resulting in an enrichment of the compound of interest. Such separators are most effective when operated at flow-rates from 1-50 cm3/min (making them useful with capillary columns), and at temperatures as high as 350°C to prevent heavy materials from condensing in the fritted glass or on other porous surfaces. The yield (Y) of this separator is in the 30-75% range, and the enrichment (N) in the 10-40% range, depending on GC column flow-rate and the pressure drop across the separator. This separator operates best at flow-rate of 20 ml/min or less.

With very polar materials, some peak distortion is
noted with this separator (mainly trailing of the
polar components into the MS). Care must be taken
when using this separator with unstable organic compounds, as some may react with active sites on the
hot glass surfaces. To overcome this difficulty,
silanization of the separator with dimethyldichlor-
osilane has been advocated(3)(usually by injection
of about 100 μl of the silanizing agent into the
separator when disconnected from the mass spectrometer inlet). Another problem often encountered
with these glass frit separators is the adsorption
and retention in the glass pores of trace amounts
of polar materials. This problem also is most readily
solved by silanization of the separator. If low vapor
pressure   materials are to be examined, the
separator and all lines leading to and from it must
be well heated. Our experience has shown that even
when operated at 250°C, when analyzing 3-4 ring
polyaromatics, many of the glass pores eventually
become clogged with organic material, resulting in
a sharp decrease in the sample yield (Y) to the MS.

Molecular Jet (Ryhage) Separator

This type of separator, constructed entirely of stainless steel, was first described by Ryhage (7). The
column effluent passes through a small orifice which
is optically aligned with another orifice. The
carrier gas (generally helium or hydrogen) diffuses
away from the line-of-sight more rapidly then the
heavier organic molecules, resulting in good enrichment of the carrier gas entering the MS with the
organic material. The molecular jet separator is
primarly used with packed columns with higher
carrier gas flows of 30-50 ml/min, 40-70% of methyl
stearate passed into the ion source of the MS (8).
The two-stage separator.The two-stage separator is
used in the LKB 9000 GC-MS tandem and numerous published results attest to its reliability. Yields (Y)
and Enrichments (N) in the range of 40-70% are
usually obtained with this type of separator. Maximum operating temperatures are near 350°C, although
the Ryhage separator is more generally operated between 200-250°C to minimize   thermal rearrangements
or cracking of thermally unstable molecules. Peak

distortion due to the separator is slight. The main
disadvantage of this type of molecular jet separator
is that it must be optimized for one column flow-
rate, and the design (i.e.-distance between jets)
must be changed for use at other flow rates. However, once optimized, the separator is rugged and
easy to operate, with very little surface interactions
with organic molecules.

One of the main items of routine maintenance with
this type of molecular separator is to make sure
the jet openings do not clog. In general, this
can be caused by condensation of high-boiling materials, and this possiblity should be taken into account when choosing the operating temperature for
the separator. If a jet opening does become plugged,
a high-temperature/vacuum bakeout will usually cure
the problem by vaporizing the organic plug. However,
if the plug is a small particle of some foreign
material that was in the carrier gas, usually disassembly and mechanical removal is required. To
prevent condensation problems, make sure the GC/sep-
arator and separator/MS transfer lines are also well
heated, with no cold spots. It is also necessary
that all vacuum connections be as leak-free as
normally required for MS work. If these few precautions
are taken, this design separator should give essentially trouble-free operation.

Silicone Membrane (Llewellyen) Separator

Llewellyn et. al. has reported use of a silicone
rubber diaphragm as a molecular separator (9). With
this separator, the eluted organic compounds diffuse
 through the diaphragm, which is relatively im-
permeable to the carrier gas (hydrogen, helium, argon,
nitrogen). Both single-stage (one membrane) and
dual-stage (two separate membranes) separators have
been used. The desired flow-rate range for the single-
stage is from 1-20 ml/min and to about 60 ml/min
for the dual-stage, with a maximum operating temperature 250°C for either type. A yield (Y) of 40-90%
and enrichment (N) of 5-40% is common for the single-
stage type, with Y=50% and N = 100 for the dual-stage
type. Although these separators are very efficient,
polar components and high-boilers tend to be partially
adsorbed on the silicone membrane, resulting in
tailing of these materials into the MS, and in some

cases, complete removal of very polar materials. One definite advantage of these separators is the GC outlet is kept at atmospheric pressure with no separate pump required. A forepump is used in the dual stage mode between Stage 1 and Stage 2 to enhance removal of carrier. Routine maintenance of these silicone membrane separators is similar to other separators in that they must be kept very clean, although the extreme high-temperature/vacuum bakeout is not possible. If the yield of the separator begins to deteriorate (i.e.-develops pinhole leaks), or if contaminants adsorbed on silicone membrane begin to gradually be emitted into the MS ion source, a fresh membrane properly conditioned by vacuum bakeout, solves these problems. It should be considered a routine maintenance procedure not to allow very polar materials to enter the separator, and thus, prevent adsorption on the silicone membrane. Other than the items mentioned, very little can go wrong with this type of separator provided the silicone membrane is intact.

Thin-Walled Teflon (Lipsky) Separator

The molecular separator system reported by Lipsky, et.al. (10) (11), is actually a modification of the Biemann-Watson separator in that the porous glass tube is replaced by about 6-8 feet of thin-walled Teflon tubing. Both ends of the Teflon tubing are connected to stainless steel capillary tubes to reduce the pressure, and allow carrier gas flow-rates of 15 ml/min or less. The Teflon tube is contained in a vacuum jacket connected to rotary pump. The optimum operating temperature range in which the Teflon was most highly selectively permeable to helium is 250-260°C above ambient, at which yields (Y) of 80-90% are obtained, but only at an enrichment (N) of 2-6%. At temperatures 200°C above ambient there was no diffusion through the membrane, and at 300°C above ambient, most of the helium and sample diffused through the Teflon (3). Thus, the use of this separator is somewhat limited.

Routine maintenance of such Lipsky separators consists mainly of keeping them exceptionally clean via high-temperature (250°C) vacuum bakeouts. In addition, they should often be vacuum checked, especially where the Teflon is connected to the metal capillaries, as leaks can easily occur here, and give

the appearance that all materials are being passed
through the Teflon in equal amounts, and thus, a
very low enrichment (N) is observed. At higher tem-
peratures(250°C+) for extended periods of time, Teflon tends to lose its elasticity, and could become
loose at the metal connections. Thus, the Teflon
membrane should be replaced often. This will also
prevent repeated leaking of heavy materials that
may be adsorbed on the Teflon into the MS ion source.
Such separators are quite rugged and should require
very little maintenance if adequate operating temperatures are maintained (250°C).

Porous Silver Frit(Blumer) Separator

A very small separator based on a porous silver frit
to selectively remove the carrier gas has been
reported by Blumer(12). This separator is mainly
for use with low-flow packed columns(10-15 ml/min).
The separator consists of a 1/4 inch stainless steel
Swagelok tee with one arm (that is connected to a
rotary pump) containing a porous silver membrane(6mm
diameter, 2 mil thickness, maximum pore diameter 3 μm).
The other two arms of the tee are connected to the
GC and the MS ion source through a needle-type
metering valve. Satisfactory analyses were obtained
with saturated and unsaturated hydrocarbons (up to
C 20) and polar aldehydes, ketones, dicarbonyl compounds and fatty acid esters (up to C22). The yield
(Y) of this separator is in the 30-40% range with an
enrichment (N) of about 10%. The maximum temperature
for operation of this separator could go as high as
400°C.

A disadvantage of this type of separator might be
silver catalyzed surface reactions of organic molecules at high temperatures; however, the separator
is very rugged, and has an inherent low dead-volume,
making it ideal for smaller diameter columns. In a
recent paper, Grayson and Levy (13) report on a Blumer-
type separator with dimensions optimized for use
with open-tubular (capillary) columns down to 0.010-
inch i.d., with flows in the 2-7 ml/min range.
Yields (Y) for this microseparator range from 25-60%
and enrichments (N) from 2-10% for various compounds
with the separator operated at 175°C and a column
flow-rate of 10 ml/min. No detectable peak broadening
or tailing was detected using this separator. A block
diagram of this separator is shown in Figure 13-4.

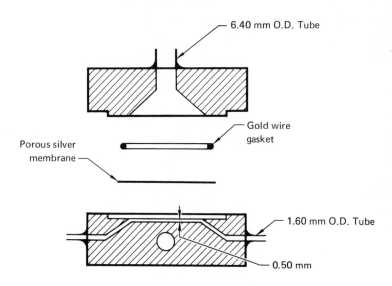

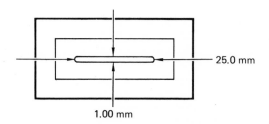

Figure 13-4 Cross section of the micro separator
(exploded view) and top view showing
dimensions of separator volume. The
6.40mm O.D. tube opening is connected
to a vacuum pump and the two 1.60 mm
O.D. tube openings are connected to the
gas chromatograph outlet and inlet
mass spectrometer. (Courtesy J.
Chromatogr. Sci.) (15)

Routine maintenance of these types of separators is
quite easy due to their simplicity of construction
and the fact they are very rugged. The major item is
once again overall separator cleanliness and membrane

cleanliness. Thin layers of organic materials on the silver frit can usually be removed by a high-temperature/vacuum bakeout as is used for other separators. If the GC side of the frit becomes covered with a thin coating of organic material when very low vapor pressure components are being analyzed, the frit should be replaced. Due to the simplicity of construction, problems with leaks should be minimal. In general, this type of separator can be operated for many months without any extensive maintenance required.

Variable Steel Plate (Brunee) Separator

The Brunee adjustable steel plate separator allows carrier gas/sample differentiation by a variable annular slit (14). The outstanding characteristic of this separator is that it may be operated over a large range of flow rates (1-60 ml/min). The yield (Y) for this separator is between 10-30% and the enrichment (N) from 10-40%, with an optimum operating temperature of about 300°C. The main disadvantage of this type of separator is possible surface reactions of organic compounds on the high temperature steel. On the other hand, the ruggedness of the separator along with its simplicity of operation and wide dynamic range make it a desirable unit to use in equipment where both packed and capillary columns will be employed.

Routine maintenance of this separator will generally consist of high-temperature/vacuum bakeouts. However, if the separator becomes clogged with any decomposed or high-boiling organic materials, it can be disassembled and all the stainless steel parts cleaned in a suitable solvent (acetone or chloroform), thoroughly dried and reassembled. Subsequent leak testing, and a high-temperature bakeout to remove any traces of the cleaning solvents is also in order.

Membrane-Frit (Grayson-Wolf) Separator

Grayson and Wolf have constructed a two-stage membrane-frit separator that is highly efficient between column flows of 5-60 ml/min (15). In addition, with this separator the GC outlet is maintained at atmospheric pressure. The yield (Y) of this separator can range from 40-60% and the enrichment (N)

from 10-40%. Maximum operating temperature for this
separator is 250°C due to the silicone membrane
portion. Peak broadening has not been a problem with
this separator.

Major maintenance items for this separator are general
overall cleanliness and checks on the condition of
the silicone membrane. If the membrane develops
pin-hole leaks (as sometimes will happen after ex-
truded operation at high-temperatures), the first
stage will not be operating as a separator at all,
and the total separator performance will deteriorate.
If the efficiency of this separator begins to decrease,
the first check point is the silicone membrane.

Conclusions Concerning Routine Maintenance and Operation of Most Molecular Separators

As shown by Grayson and Wolf (4) and discussed in the
previous section, several GC factors can affect separator operation, and as a result, affect MS resolution.
The most prominant are: 1. Carrier flow-rate--the
separator must be operated in the correct range,
2. Carrier/sample ratio--if the separator is not
being operated in its area of greatest efficiency, a
high ratio will exist which generally has a detrimental
effect on MS resolution and sensitivity, 3. Peak
elution width--if a peak is too sharp, the rapidly
changing sample pressure will cause mass spectral
resolution anomalies. On the other hand, if the peak
is too wide, it is almost impossible to obtain a
representative spectrum due to lower than optimum
sample/carrier gas ratio and 4. Impure peaks containing more than one component--they may give excellent
MS resolution but a confusing spectrum. The experienced operator can usually spot a mixture from its
mass spectrum. However, probably the most important
GC condition to be maintained for proper separator
performance is to keep the column flow-rate (Item 1)
well within the limits prescribed for any given
separator.

Generally, for a good separator performance, the unit
should be operated within the specification given for
the individual separators in the previous section of
this chapter; otherwise, poor overall performance will

result. It should be stressed that the separator and all connecting tubing and valves should be heated to the maximum extent the components in the various samples will allow, and these temperatures should be checked periodically to detect any possible cold spots.

In our experience, capillary columns have been operated such that 50-90% of the effluent goes to the molecular separator/MS system while the remaining 10-50% is directed to a GC detector (usually an FID). Thus, we obtain the standard chromatogram along with mass spectra of desired peaks. Some newer units rely on the total ionizing current in the MS for the chromatograph, and the total column effluent is directed to the separator. However, one must be careful when directing the total column flow to the separator not to exceed the optimum flow-rate range of the separator, and install a split valve if necessary (especially with packed columns). An obvious malfunction that can be devastating to the performance of a separator system is a leak somewhere between the column exit and the MS ionization source. If a leak is present, air and other contaminants will be sucked in and pass into the MS ionization region resulting in high background level. Also, to keep the background level down after a leak has been tracked down and terminated the separator system should be baked out at a high temperature under vacuum as described previously. Several good high-temperature vacuum leak sealants are commercially available, with the aerosol variety being particularly effective in stopping leaks found in the tube fittings used with most GC-separator-MS installations. The best procedure is simply to spray a given joint with sealant, and observe the mass spectrum for declining intensity of the water series (m/e 16-18), nitrogen (m/e = 28), oxygen (m/e - 32), argon (m/e=40), and CO_2 (m/e=44).

These are only a few suggestions to keep the GC/MS system operating at top efficiency. The items to remember to keep a separator system operating most effectively are:
1. Do not exceed the specified flow-rates for the separator.
2. Keep it as warm as possible.
3. Keep it clean by periodic vacuum bakeouts.

4. Be sure and use the correct system for the type of compounds being analyzed to prevent loss or sample decomposition (thermal or metal catalyzed).
5. Make sure the system is leak-tight, as "air pollution" of the GC effluent can cause high MS background.

If these general rules are followed, the mass spectrometer should yield invaluable compositional data for separated components on a routine basis.

Thus, in this chapter we have discussed some of the more common auxillary techniques used for qualitatively monitoring GC column effluents (other then the conventional GC detectors), and have endeavored to point-out some routine maintenance procedures that should aid in the optimum operation of these auxillary systems.

References

1. J.T. Walsh and C. Merritt, Anal. Chem., 32, 1378 (1960).

2. R.S. Gohlke, Anal. Chem., 31, 535 (1959).

3. D.I. Rees, Talanta, 16, 903 (February 1969).

4. M.A. Grayson and C.J. Wolf, Anal. Chem., 39, 1438 (1967).

5. J.T. Watson and K. Biemann, Anal. Chem., 36, 1135 (1964).

6. ibid., 37, 844 (1965).

7. R. Ryhage, Anal. Chem., 36, 759 (1964).

8. R. Ryhage, Arkiv. Kemi., 26, 305 (1967).

9. P. Llewellyn and D. Littlejohn, Varian Technical Quarterly, p. 6 (Spring 1966).

10. S.R. Lipsky, C.G. Horvath, and W,J, McMurray, Anal. Chem., 38, 1585 (1966).

11. S.R. Lipsky, C.G. Horvath, and W.J. McMurray,
    Gas Cromatog., 1966 ed., A.B. Littlewood, p. 299.

12. M. Blumer, Anal. Chem., 40, 1590 (1968).

13. M.A. Grayson and R.L. Levy, J. Chrom. Sci.,
    9, 689 (1971).

14. C. Brunee, et. al., 17th Annual Conference on
    Mass Spectrometry and Allied Topics, Dallas,
    Texas (May 1969), Paper no. 46.

15. M.A. Grayson and C.J. Wolf, Anal. Chem., 42,
    426 (1970).

Chapter 14

GAS CHROMATOGRAPHIC COMPREHENSIVE TROUBLESHOOTING

Introduction

The most expensive GC instrument is worthless if it does not operate properly. Most instruments are so complex and so many variables are involved that it may not be easy to determine the cause of the difficulty. Short of simple luck, the best method of attacking the problem is a logical one in which each possible cause is eliminated in the order in which they are most likely to occur. In the following pages GC troubleshooting is brought into a simple and quick form.

The tables are divided into several categories descriptive of the symptoms found frequently in ailing chromatographs. Each category is followed by three sections designed to pinpoint the problem and find a remedy. The first section defines the symptom and, where pertinent, provides sample chromatograms to illustrate the problem. Following this section is a section entitled "Accompanying Symptoms." Often more than one symptom is present providing a clue as to the origin of the problem which is causing these symptoms. In this table, the entire GC system is broken down into four divisions as follows:

1. Gas System (Gas): The gas system, here, will include all GC parts involved in moving the sample through the column. The tubing, gas supply, flowmeters, etc. are included.

2. Column (Col): The column includes the column tubing, its packing and/or liquid phase, as well as the injector or sampling values.

3. Detector (Det): The detector is considered exclusive of the electrometer, bridge, and power supply. Thermal Conductivity Detectors (TCD),

Flame Ionization Detectors, (FID), and Electron Capture Detectors (ECD) are considered in particular. Included with the detector are GC parts involved in transporting detector gases (H_2, air) to the detector.

4. Electronics (Elec): Electrometers, power supplies, bridges, recorders, and oven and detector heaters are included here.

If an accompanying symptom is present, the most likely system to troubleshoot and the order in which to troubleshoot the other systems is given.

In the third section a table is given which lists the probable causes of the problem indicated by the symptoms. Where, as in drift, the chromatogram may take on more than one shape, Column 1 indicates the type of chromatogram evidenced. Column 2 lists the system where the problem originates. Causes of the symptom are listed in the next column followed by methods of isolating and confirming the source of the difficulty. Means by which the problem can be remedied are given in the next column, through the remedy is often obvious once the problem is isolated. In the final column, a reference page or pages is given to direct the reader to a more detailed analysis of the subject.

In the trouble-shooting tables, the systems are dealt with in the order in which they occur in the GC instrument. The most likely cause of the problem is dependent upon many variables such as the type of sample being analyzed, the type of column being used, and, perhaps most important, the individual "quirks" of the instrument.

Baseline Problems: Drift

A. Rising Baseline   Falling Baseline

(Temperature Programming)

B. Rising Baseline   Falling Baseline

(Isothermal Operation)

C. Irregular Drift

228

Definition

Drift is characterized by changes in the baseline. Drift is generally made up of slow fluctuations in the baseline when no sample has been injected. The baseline may either rise or fall steadily as in A and B (page 2) or it may fluctuate as in C (page 2).

Drift: Accompanying Symptoms

Accompanying Symptoms	Order in which to Trouble-shoot Systems			
	Gas Flow	Col-umn	Detec-tor	Elec-tronic
Poor Retention Time Reproducibility	1	2	--	--
Noise	2	1	3	4
Noise with Sensi-tivity Loss	1	--	2	3
Drift Increases or Decreases with Col-umn Temperature	2	1	3	--
Drift Decreases with Time	2	1	3	--
Detector Tempera-ture Fluctuating (TCD)	--	--	1	--
Drift at Attenuation	--	--	--	1

Baseline: Drift Problems

Drift Type	System	Cause	Isolate and Confirm	Remedy
A or C	Col	Column bleed	Drift decreases with temperature decrease	Recondition column at near the maximum liquid phase temperature for one hour or until drift disappears (disconnect column from detector), Chapter 10. Use dual column system (see Chapter 10, columns).
	Col	Contamination of column		Recondition and/or replace column, Chapter 10.
B	Col	Unstable column oven	Monitor oven temperature	Check for proper oven fan operation and air flow, Chapter 10.
				Replace temperature controller, Chapter 10.
	Elec	Defective electrometer	Disconnect detector cables	If drift continues, replace electrometer, Chapter 12.
			Check electrometer batteries	Replace batteries every three months, Chapter 12.

Baseline: Drift Problems (cont.)

Drift Type	System	Cause	Isolate and Confirm	Remedy
B	Elec	Unstable TCD detector temperature	Monitor detector temperature	Replace temperature controller, Chapter 11.
C	Gas	Poor H_2 gas regulation (FID)	Check H_2 flow at detector for proper and constant flow	Readjust flow or replace flow controller, Chapter 8. Leak check Chapter 8.
C	Gas	Carrier gas leak	Leak check all fittings	Tighten or replace fittings
C		Poor carrier gas regulation	Monitor gas flow at column outlet	Be sure supply tank is maintained at a sufficient pressure to ensure proper operation of flow controllers
				Repair or replace flow controllers
C	Det	Poor instrument location	Stop room air condition system and check drift	Move instrument away from heating and cooling outlets or other drafts which might affect temperature of the instrument environment

231

Baseline: Drift Problems (cont.)

Drift Type	System	Cause	Isolate and Confirm	Remedy
C	Det	Detector contamination	Reduction in detector temperature should decrease drift	Clean detector according to Appendix A, Detector Cleaning
C	Elec	Instrument not grounded	May be accompanied by noise	Ensure instrument and recorder securely attached to a good earth ground
	Elec	Electrometer defective (FID,ECD)	Isolate electrometer from detector to see if drift continues	Check zero circuitry, batteries balance circuit
	Elec	Electrometer not warmed up	Drift will discontinue when instrument has had time to warm up	Warm up- time may require as long as 24 hours
	Elec	Bad bridge (TCD)	Movement of bridge wires increases or decreases drift	Check power supply, zero pots, etc.
D	Gas	Carrier gas supply pressure too low	Monitor flow at outlet of column	Increase carrier gas supply to pressure regulator and flow controllers
	Col	Unsteady column oven temperature	Check oven seals	Make sure oven is sealed properly and proper oven insulation is in place

Baseline: Noise

Noise Type	System	Cause	Isolate and Confirm	Remedy
A	Gas	Dirty injector	Low injector temperature should decrease noise	Clean injector and replace septum
		Leak	Leak Check at all connections	Tighten or replace connections
		Carrier gas flow too high	Check flow at column outlet	Readjust flow controllers to obtain optimum flow rate
		Contaminated carrier gas	Replace carrier gas supply	Discard contaminated carrier gas
A	Col	Contaminated column	Noise should decrease at lower temperatures	Recondition column for at least one hour at slightly less than the maximum temperature of the liquid phase. It may be necessary to replace a badly contaminated column
		Column bleed	Noise should decrease at lower temperatures	

233

Baseline: Noise (cont.)

Noise Type	System	Cause	Isolate and Confirm	Remedy
A	Det	Faulty detector cables	Disconnect cable at electrometer. If detector has checked out O.K. and noise discontinues, cable is at fault	Replace cables
		Contaminated air or hydrogen (FID)	Replace hydrogen and air supply	Discard contaminated gas supplies
	Det	Improper hydrogen flow (FID)	Check hydrogen flow at flame tip	Adjust hydrogen flow as indicated in the instrument manual
		Improper air flow (FID)	Check for proper air flow rate	Adjust air flow to proper level
		Water condensing inside detector (FID)	Check detector temperature	Raise detector temperature to slightly higher than 100°C
		Contaminated detector	Lower detector temperature (but not below 100°C) noise should decrease	Clean detector as outlined in Detectors Chapter

234

Baseline: Noise (cont.)

Noise Type	System	Cause	Isolate and Confirm	Remedy
A	Det	Dirty detector insulators (ionization detectors)	Noise should decrease at lower temperatures	Clean with a residue-free solvent. Do not handle clean insulators
A or B	Elec	Loose cable connections	Check all plugs and screw connections to make sure they are solid	If necessary, clean or replace bad connectors
	Elec	Malfunctioning integrator feedback	Bypass integrator in system	Change cable connections or consult instrument manual
	Col	Column packing entering detector	Change column	Increase flow to remove loose particles. Apply back pressure through detector (disconnect column)
A or C	Elec	Bad ground	Check to see that ground connection is secure	Be sure that the ground connection is attached to a good earth ground
		Dirty switches	Examine switches for heavy oxidation and firm seating	Clean switch contacts with either a light grit emery paper or contact cleaner

Baseline: Noise (cont.)

Noise Type A or C	System	Cause	Isolate and Confirm	Remedy
C	Elec	Dirty recorder slide-wire	Noise will probably be intermittent, occurring always at the same point on the recorder scale. Noise will not change with attenuation	Clean slide-wire according to recorder manual. Special slide-wire cleaner is available for this purpose
		Recorder not operating properly	Short recorder input. If noise continues, recorder is source	Check recorder damping and gain control. Consult recorder manual for further troubleshooting
C	Gas	Hydrogen generator H_2O overflow	Check hydrogen generator for H_2O overflow	Dry transfer lines from generator to column with heat or replace
C	Elec	AC input circuit overload	Remove other loads from circuit into which GC is hooked	Reduce circuit load or change AC input circuit for GC

236

Baseline Problems: Noise

Types of Noise:
A. Irregular or Intermittent Noise

B. Regular Noise

C. Spikes

Definition:

Noise is characterized by a jerky recorder pen. At
times, as in Type A (above), the noise is erratic and
may disappear completely for short periods of time.
A second type of noise, shown in B, is a regular
pattern of spikes. Noise is not necessarily caused
by the recorder or any other segment of the electrical
system, and often it is caused by contamination of
the column or gas system.

Accompanying Symptoms

These additional symptoms may provide a guide or
direction to the origin of the problem. The system
in which the problem originates is listed in the
troubleshooting tables which follow. The most logical order of troubleshooting is indicated.

Accompanying Symptom	Gas	Col	Det	Elec	Remarks
Drift (less at lower temperature)	2	1	3	--	Most likely column bleed
Noise and Drift De-crease with Time	3	1	2	--	Contamination
Sensitivity Loss	--	--	1	2	Wrong flow with ionization detectors
Poor Retention Time Reproducibility	1	2	--	--	Common with temperature programming
Drift at Attenuation (Short	--	--	--	1	Check out recorder

Baseline Problems: Zeroing

This problem arises when the recorder cannot be set at zero. The cause of zeroing difficulties are varied; and, while it is not a frequent problem, zeroing difficulties may be the most difficult to remedy.

Baseline: Zeroing

System	Cause	Isolate and Confirm	Remedy
Gas	Improper flows (carrier, air, H_2)	Check flows with bubble meter	Reset flows according to instrument manual
Col	Excessive column bleed	Lower column temperature, zero should go down	Recondition column at near maximum liquid phase temperature until bleed stops
Det	Dirty detector (particularly FID and ECD)	Disconnect cables at detector and attempt to zero	Clean as outlined in Detectors Chapter
	Weak filament (TCD)	Disconnect cables at detector and attempt to zero	Replace filaments
Elec	Recorder zero not set	Short recorder input and check zero	With input still shorted, set recorder zero
Elec	Malfunctioning electrometer		Consult electrometer manual
	Improper recorder connection	Check all recorder connections for proper positioning	Connect cables to recorder as outlined in instrument manual
	Defective recorder	Short recorder input and check zero	If zeroing not possible, consult instrument manual for troubleshooting
	Bad battery	Check with voltage meter	Replace batteries

Baseline Problems: Cycling

Cycling is a baseline which shows a regular wave pattern. The problem is most likely caused by fluctuations in the carrier gas flow rate. In the case of TCD detectors, the problem can be caused by variations in the detector cell temperature.

Baseline Problems: Cycling

System	Cause	Isolate and Confirm	Remedy
Gas	Carrier flow variation	Monitor flow at detector	Flow controllers require high inlet pressures (30 psi or above). Replace faulty flow controllers
	Pumped gas supply	At high sensitivities, variations in supply pressure from pumped gas supplies will cause cycling	Use a ballast tank to eliminate pressure variations
Det	Cycling detector temperature	Monitor detector temperature. Check line voltage supplying controller instrument	Repair or replace faulty detector temperature supplying controller
	Condensation in TCD detector		Increase flow rate through detector
			If necessary, clean detector as outlined in Chapter on Detectors

Baseline Problems: Spikes

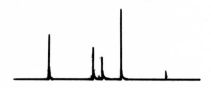

Unlike noise, spikes often travel nearly completely
across the recorder range. Spikes are generally
intermittent and vary in size. Usually caused by
dirty electrical switches and connections or a
contaminated detector, spiking may also result from
line voltage fluctuations or drafts.

Baseline: Spikes

System	Cause	Isolate and Confirm	Remedy
Gas	Contaminated gas	Connect gas supply directly to detector	Replace gas supply
Col	Loose column packing	Tap column sharply and observe baseline	Disconnect from detector and increase flow rate
Det	Drafts	Check for drafts from air conditioners, blowers, opening doors	Move instrument
Det	Dirty venting tube (ECD)		Clean as outlined in Chapter on detectors
Elec	Dirty electrical connections (FID, ECD)	Spikes will continue with electrometer in "Balance" position	Clean connections, switches, pots and troubleshoot power supply
Elec	Zeroing circuit	Spikes will stop with electrometer in "Balance" position	Consult instrument manual for troubleshooting zero circuitry
Elec	Dust, dirt in detector (FID)	See remedy	Blow a clean, dry air supply across detector
Elec	Line voltage fluctuations	Monitor line voltage coming into instrument	
Elec	Vibrations	Usually aggravates poor electrical connections	Remove source of vibrations (i.e., oven fan) or isolate instrument from shock

Distorted Peaks: Ghost Peaks

Types of Ghost Peaks

A.   Ghost Peak

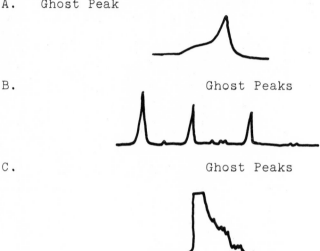

B.   Ghost Peaks

C.   Ghost Peaks

Often during the course of an analysis, peaks will
appear which do not fit the expected pattern of the
chromatogram. These unexpected peaks, known as
Ghost Peaks, are many times not well resolved (A),
are generally present in small quantities (B), and
sometimes resemble noise as they emerge together
with the larger peaks of the chromatogram (C). The
cause of these ghost peaks is most often found in
the inlet system, the gas system, or the column
system.

Ghost Peaks

Type	System	Cause	Isolate and Confirm	Remedy
B,C	Gas	Contaminated sample	Obtain another sample be sure to clean syringe thoroughly	Use purest available samples
A,B,C,	Gas	Leak	Leak check all fittings at maximum temperature at which they are used	Tighten or replace worn fittings
A,B,C,	Col Gas	Contamination	Lower system temperatures and observe effects	Clean or replace contaminated system components
		Septum bleed	Note maximum recommended temperature for septum. Cover new septum with aluminum foil	Pre-condition septums at operating temperature Replace worn septums
A,B,C	Col	Reaction of sample with column liquid phase	Inject a different type of sample	Use a different column
		Reaction of sample with other GC parts (Septum, column tubing, fittings, etc.	If sample composition is known, examine reactivity with GC system components	Where possible, use less reactive parts made of quartz glass, Teflon, stainless steel

243

Ghost Peaks (cont.)

Type	System	Cause	Isolate and Confirm	Remedy
A	Col	Heavy M.W. material from previous analysis is eluting		Condition column at near maximum temperature for about one hour or until a steady baseline is observed
	Col	Condensed materials (H_2O or other impurities) eluting during temperature programming		Install or replace carrier gas filter
B	Gas	Air peak (not present with FID)	Air peak is usually the first peak to emerge	Cannot be eliminated with making syringe injections, if necessary, air peak can be eliminated by using a sampling valve
B	Col	Column liquid phase desorption when injecting a solvent	Inject a sample which does not include solvent	Inject solvent into column repeatedly and recondition column
C	Col	Decomposition of sample	Lower injector temperature	Lower injector temperature (if high enough to cause thermal decomposition) Use different column liquid phase (if phase is not compatible with sample)

Distorted Peaks: Others

Types of Distorted Peaks

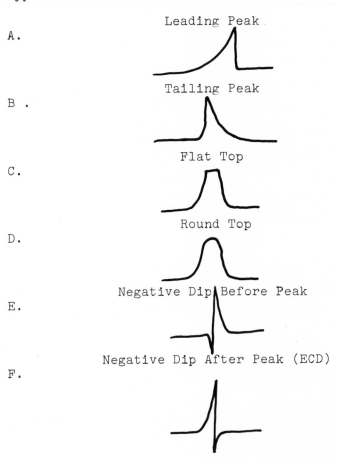

A.

Leading Peak

B .

Tailing Peak

C.

Flat Top

D.

Round Top

E.

Negative Dip Before Peak

F.

Negative Dip After Peak (ECD)

The ideal chromatogram contains all very nearly
Gaussian shaped peaks. Other peak shapes are often
encountered caused usually, by improper operating
conditions. Column overload is the most common
cause of almost all types of distorted peaks.

245

Distorted Peaks

Type	System	Cause	Isolate and Confirm	Remedy
A	Col	Column overlaod	Inject smaller sample	Use smaller sample
				Use larger column
		Sample condensation	Check vapor temperature of sample	Use high enough injector and column temperature
		Two peaks eluting simultaneously	Lower column temperature about 30°C to try for better separation	Optimize column operating conditions
				Use a more suitable column
	Gas	Poor sample injection	Inject another sample	Use proper injection techniques as outlines in Appendix B
		Flow through detector too slow	Check flow at detector outlet. Carrier flow of at least 30 ml/min should be entering detector	Use scavenger gases at end of column where necessary to increase flow of sample into detector
	Col	Injector temperature too low	Raise injector temperature	Use high enough injector temperature to readily volatilize the sample

Distorted Peaks (cont.)

Type	System	Cause	Isolate and Confirm	Remedy
A	Col	Interaction with column support material	See remedy	Use a less active support
				Increase liquid loading of column
				Derivatize sample
		Column over-load (with gas samples)	Inject smaller sample	Use smaller sample
				Use larger column
		Dirty injector tube	Check for sample residue. Increased injector temperature	Clean injector tube by flushing with solvents.
B	Col	Two compounds eluting to-gether	Lower column tempera-ture to improve separation	Optimize column operating conditions (flow, tempera-ture)
				Use different column (liquid phase diameter, length)
		Oven tmepera-ture too low	Increase oven temperature	Use higher oven temperature if resolution remains good
C	Elec	Electrometer saturated	Switch electrometer	Use smaller sample
				Use different electrometer range

Distorted Peaks (cont.)

Type	System	Cause	Isolate and Confirm	Remedy
C	Elec	Recorder range exceeded	Switch recorder range	Use smaller sample Use different recorder range
D	Det	Operating outside of linear range of detector (most likely with ECD)	Examine linear range of the detector in use	Reduce sample size
	Elec	Recorder gain set too low	See remedy	Adjust recorder gain to proper setting
	Gas	Pressure surge	Often seen before large peaks	Not easily eliminated: try 1. Reduce sample size 2. Reduce column bleed 3. Check for contaminated detector
F	Det	ECD sample overload	See remedy	Allow time for baseline to return to normal dilute sample or reduce size
		ECD, dirty detector	See remedy	Clean detector as outlined in the Chapter on detectors

Reproducibility Problems: Retention Time

Most uses of chromatography require that the retention
times of peaks be reproducible. Non-reproducibility
may develop suddenly or come about as a result of
gradual changes. In isothermal GC, the most common
cause of retention time difficulty lies within the
flow system and followed closely by the column system.
In temperature programmed GC, by far the most common
cause of differing retention times is non-reproducible
temperature profiles.

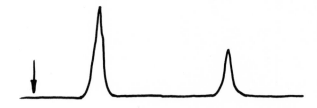

Accompanying Symptoms

These additional symptoms may provide a guide to the
origin of the problem.

		System			
Accompanying Symptom	Gas	Col	Det	Elec	Remarks
Noise	1	2	--	--	Probably a leak in carrier gas
Sensitivity	1	--	--	--	Probably a leak in carrier gas
Drift	1	3	--	--	
Distorted Peak	1	2	--	--	Sample size too large is another possibility
Gradual Decrease in Retention Time	2	1	--	--	

Reproducibility: Retention Time

System	Cause	Isolate and Confirm	Remedy
Gas	Insufficient retention	Peak measured should require at least four times as long to elute as does the "air" peak	Lower temperature of column Lower flow rate through the column
	Poor injection	Change method of syringe handling and/or syringes	Syringe injection should be accomplished quickly and cleanly as detailed in Appendix B
	Leak	Replace septums	Septums should be replaced periodically; especially during high temperature operation or frequent injections Leak check all fittings
	Changing flow (while temperature programming)	Check flow at column outlet at both minimum and maximum temperature	Flow rates should not differ by more than 2 cm3/min for a 1/8 in. I.D. and larger columns
	Bad flow control	Monitor flow rate at column outlet	Increase flow controller inlet pressure
Col	Column temperature not equilibrated		Allow five minutes for temperature to equilibrate after reaching operating temperature

Reproducibility: Retention Time (cont.)

System	Cause		Remedy
Col	Column temperature not equilibrated	Isolate and Confirm	When temperature programming, allow time to equilibrate at starting temperature. More time is required when starting near ambient
Col	Bad column temperature control	Check oven seal	Make sure oven is seated properly before beginning
		Monitor column temperature	Troubleshoot temperature programmer (gears, and thermal couple)
	Too much sample	Peak tailing will result	Dilute sample or reduce amount injected
			Go to larger column
	Operation outside of range of column liquid	Consult Appendix C for proper range	Generally, do not operate within 5°C of either minimum or maximum temperature
	Column Deterioration	Probably accompanied by loss of resolution	Replace column

Reproducibility Problems: Sensitivity Changes

Within the limits of injection techniques, the relative peak areas should be the same in successive
analysis. Most changes in sensitivity are actually
changes in the amount of sample introduced into the
column brought about by improper injection techniques.
Even with injection techniques optimized, absolute peak a
peak area will vary by at best around 5%. Another
frequent cause of sensitivity change (loss) is a
leak in the gas system. Changes in carrier gas flow
will be inversely proportional to the sensitivity
of ECD and TCD detectors.

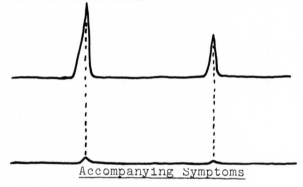

Accompanying Symptoms

Accompanying Symptom	Suspected System Gas	Col	Det	Elec	Remarks
Drift	1	2	3	--	Leak check
Noise	1	2	--	--	Leak check
Baseline Not Return-ing to Zero	--	--	2	1	Check recorder zero
Ghost Peaks	2	--	1	--	Bleed or con-tamination
Round Top Peaks	--	1	--	--	Overload detector
Changes in Re-tention Time	1	--	3	2	Changes in carrier gas flow with ECD, TCD

Sensitivity Changes

System	Cause	Isolate and Confirm	Remedy
Gas	Injection technique	Repeat analysis. Frequent rising and falling of peak area indicates poor injection techniques	Proficiency in the use of a syringe requires practice. See Chapter for proper methods of injection. Maximum repeatability is about 5% with the best syringes
	Carrier gas flow changes (ECD, TCD)	Check carrier gas flows	Readjust flow or replace flow controller if needed
	Changes in H_2 flow (FID)	Check H_2 flow at flame tip	Optimize H_2 flow (normally in approximately 1 to 8 ratio H_2 to air)
Gas	Leak	Measure all flows at the detector and leak check if necessary	Tighten or replace connections
Col	Faulty syringe	Use a different syringe if available	Replace syringe needle, plunger, seals, etc: as needed or return to manufacturer for repairs

Sensitivity Changes (cont.)

System	Cause	Confirm	Remedy
Col	Incon-sistent sample prepara-tion	Isolate and Review sample preparation procedures for possible inconsistencies	Use as few steps as possible in preparing sample Hold all conditions constant Take extreme care to avoid sample contamination from syringes, sample vials, atmosphere, etc.
Det	Improper collector alignment	Check alignment of collector with respect to flame jet and igniter	Consult instrument manual for proper alignment
	Dirty de-tector (FID)	Probably accompanied by noise	Clean as outlined in Chapter 11
	Dirty filaments (TCD)	Accompanied by increased zero	Clean detector as outlined in Chapter 11
	Short in cell (ECD)	Check anode lead to ground voltage. It should read 2-4 volts. If less, disconnect cable and recheck at the connector. If now normal, short is indicated. If voltage still low, trouble-shoot pulser	Replace cable and examine detector for possible shorts

254

Sensitivity Changes (cont.)

System	Cause	Confirm	Remedy
Det	Detector overload	Isolate and Confirm Rounded top peaks will probably be evidenced	Stay with linearity of detector. Reduce sample size
Elec	Varying column temperature	Will be accompanied by changes in retention time	Repair or replace temperature controller or programmer
	Recorder settings changed	Check recorder damping and gain	Readjust recorder damping and gain
Elec	Low collector electrode voltage (FID)	Extinguish flame, measure collector voltage	See instrument manual for appropriate collector electrode voltage
		Check battery voltage	Replace batteries
	Faulty pulser (ECD)	A. Zero recorder pen at 5 usec pulse interval and sensitivity of 100 X 8 (tritium) or 100 X 8 (nickel) B. Slowly increase pulse interval to 150 usec and return C. Pen should reach midway across recorder and return D. If not, pulser is faulty	Consult instrument manual for repair procedure

255

Other Problems

Not all problems are connected to the chromatogram. Some difficulties are evident just by checking currents and temperatures of various systems within the instrument. These problems can occur in heaters, detector filaments, attenuators, etc. There is little difficulty with isolation and confirmation with these problems, and the remedies are generally obvious. For these reasons, we shall depart from the previous format and simply list the problems together with their probably causes.

A. Oven won't heat (detector or column)

  1. Defective switch

  2. Loose wiring

  3. Defective powerstat

  4. Blown fuse

  5. De ective uppor limit switch

  6. Open heater element

  7. Open programmer interconnect

B. Injector or collector won't heat

  1. Heater element open

  2. Powerstat or controller defective

  3. Broken wire

C. Heater on but pyrometer will not indicate temperature

  1. Defective thermocouple wire or connection

  2. Defective selector switch

  3. Defective pyrometer (if all selections show no temperature)

Other Problems (cont.)

D. Flame goes out (FID)

　　1. Restriction in flame tip

　　2. Loss of hydrogen supply

　　3. Loss of air supply

　　4. Carrier gas flow too high

　　5. Sample too large

E. Igniter will not ignite (FID)

　　1. Broken igniter coil

　　2. Defective igniter voltage cable

　　3. No voltage out of the electrometer

F. No filament current (TCD only)

　　1. No output from power supply

　　2. Defective switch

　　3. Blown fuse in power supply section

　　4. Open current meter

　　5. Open wiring

　　6. Filaments burned out

G. Won't attenuate properly (all detector)

　　1. If faulty at only one step,
　　　　a. Check resistor for the step

　　2. If faulty at all steps,
　　　　a. Check recorder electrical zero
　　　　b. Check resistor common to all steps
　　　　c. Check recorder hookup

One of the most disappointing occurences in GC is injecting a carefully prepared sample and waiting patiently for the peaks to emerge only to have none appear. The most common causes of no peaks are the detector power being off, the flame not being ignited (FID), the carrier gas leaking, and a bad syringe. This problem is also related to "Reproducibility Problems: Sensitivity Changes" so conjuct that section also.

No Peaks

System	Cause	Isolate and Confirm	Remedy
Gas	No Carrier Gas Flow	Check tank pressure and column outlet flow	Turn on carrier gas flow and set to proper flow rate
	Leak	Leak check all fittings. Some small peaks may show up	Tighten or replace defective fittings.
	Leaking Septum	Replace setpum	Change septums frequently depending on the injector temperature
	Hypodermic Syringe Defective	Check syringe deliverly into a piece of dry tissue paper	Repair or replace syringe
Det	Cold Injector Flame not Ignited (FLD)	Check injector temperature. Check recorder zero or check for H2O condensate above flame	Increase temperature of injector. Re-ignite flame. Cleaning of ignitor may be necessary. Also Check H_2 flows at flame tip

258

No Peaks (cont.)

System	Cause	Isolate and Confirm Check Switches	Remedy
Det	No Detector Power		Turn detector switches to on position
	No Cell Voltage being Applied to Detector (ionization detectors)	Monitor cell voltage if possible. Check switches	Turn switch on. Trouble-shoot power supply
Elec	Recorder Improperly Connected	Check recorder cables for proper orientation	Hook recorder cables to proper position
	Recorder or Electrometer Attenuated too High	Increase attenuation and check for peaks	Operate recorder and/or electrometer at a more sensitive position
	Recorder Defective	If recorder pin can be moved easily by hand recorder is defective	Consult instrument manual

Appendix A

Cleaning of GC Detectors

During the use of the GC instrument, the detector
may become contaminated due to column bleed or sample
residues. Whenever possible the detector should be
cleaned simply by heating the detector block near
the maximum temperature limit for the detector.
Wrapping glass or asbestos insulated heater tape
around hard to heat areas may aid in cleaning the
detector and prevent future build-up of contamination.
Caution: Never heat detectors containing radioactive sources above limits set by the Atomic Energy
Commission. Excessive heating may also damage teflon
insulators.

If heating fails to remove contamination it will be
necessary to remove the detector and clean it using
suitable solvents. An ultra-sonic cleaner will be
helpful in cleaning many parts of the detector, but
generally a few common reagents are sufficient to
clean them.

Procedures are given here for cleaning thermal conductivity detectors, flame ionization detectors, and
electron capture detectors. These procedures may
be applicable to other similar detectors, but care
should be taken to insure that the cleaning solvents
do not damage detector parts. It is important that
cleaned detector parts not be handled with the fin-
gers. Lint-free linen gloves and tweezers should
be used to prevent recontamination of the detector.
After cleaning, the detector should be returned to
the GC instument and maintained at operating temperatures overnight before use.

Thermal Conductivity Detectors

The following procedure should be used when cleaning
the TCD detector:

1. Disconnect all electrical connections except
   detector block heater.

2. Cap off the outlet port of the detector and fill
   the detector with Decalin   (Analabs Inc.,
   New Haven, Conn.) through the inlet port.

3. Allow the Decalin to remain in the detector at 100°C for 15 min., then drain.

4. Repeat this procedure three times or until the Decalin comes out clear.

5. Substitute Dimethyl Formamide for the Decalin and repeat steps 2,3,and 4.

6. Repeat steps 1,2,and 3 using Methanol at 60°C.

7. Repeat steps 1,2,and 3 using water at 95°C.

8. Repeat steps 1,2,and 3 using Acetone at 55°C.

9. Repeat the entire cycle if necessary to completely clean the detector.

10. Pass carrier gas through the detector for about 20 min. after draining it of all solvents and before elevating the detector to operating temperatures.

11. Allow detector to remain at operating temperatures overnight before using.

Flame Ionization Detectors

The following procedures are recommended for cleaning flame ionization detectors:

A. Cleaning detector while in place in the GC instrument(for slight contamination).

    1. Remove separation column and insert clean tubing between injection port and detector.

    2. Maintain detector temperature and column oven above 125°C.

    3. Inject three 10μl samples of distilled water.

    4. Inject three 10μl samples of Freon 113(TM, Union Carbide, Cleveland, Ohio).

    5. Allow detector to remain at 125°C for one hour and check for baseline stability.

6. If contamination is still present, repeat the procedure or disassemble detector for cleaning as outlined below.

B. Cleaning of disassembled detector (for heavy contamination.)

1. Turn off hydrogen and air supplies and disconnect all electrical connections.

2. Cool detector heater. (Cooling time can be shortened by blowing cool air directly on the detector.)

3. Cap off hydrogen and air inlets.

4. Remove upper flame assembly: barrel, electrodes, and flame jet. (Consult instrument manual for proper disassembly procedure.)

5. If flame jet is made of quartz, clean it in an aqua-regia solution, If the flame jet is made of stainless steel, polish it with # 500 emery paper.

6. Wash upper flame assembly(including jet) three times in an ultrasonic cleaner using a 50:50 volume mixture of methanol and benzene(Caution: vapors are both toxic and flammable.)

7. Finally, clean these ports in pure methanol and dry in an oven at 80°C(Be sure to avoid recontamination during drying.)

Note: Do not allow halogenated solvents to come into contact with the thermal insulators(teflon, TM) since these will react causing noise to develop upon reassembly.

8. The transfer lines between the column and the flame jet can be cleaned by passing a 50:50 volume mixture of methanol and benzene through them. Continue flushing the transfer lines until the solvent emerges clear, then flush again.

Note: The methanol/benzene solvent combination

is effective for removing most common contaminants. Use of unusual samples or
liquid phases may indicate other solvents
might be more effective to remove the
contamination.

9. Reassemble the detector, being careful to
    avoid recontamination, and pass dry carrier
    gas through it for at least twenty-minutes
    before attempting to heat the detector.

10. Before igniting the flame, allow the detector
    to remain at operating temperatures(125°C
    or above) for several hours.

Electron Capture Detector

Upon obtaining the necessary AEC authorization, the
ECD can be cleaned as follows:

1. Carefully disassemble the detector, handling
    the radioactive foil(H^3 or Ni^{63}) with for-
    ceps and gloves.

2. Clean metal and teflon detector parts(ex-
    cluding the radioactive foil) in a solution
    of 2 parts H_2SO_4, 1 part HNO_3 , and 4 parts
    H_2O. Use an ultrasonic cleaner if available.
    Repeat cleaning until solutions are clear.

3. Rinse all parts thoroughly and repeatably in
    distilled water, then acetone.

4. Dry in a column oven at 80°C.

5. Specifically authorized licensees may clean
    the foil as follows:

6. Tritium (H3) Foil can be rinsed in hexane
    or pentane but never in water. The solvent
    should be discarded with large amounts of
    water (consult AEC rules and regulations
    title 10, CRF 20).

7. Nickel (Ni63) Foil should only be handled
    with six inch or longer blunt forceps and
    should never come into contact with the skin.
    Either, ethyl acetate rinsed with sodium

carbonate, or benzene can be used as solvents. The foil should then be immersed in boiling water for five minutes. While AEC regulations should once again be consulted, these solvents can usually be rinsed down the drain with large amounts of water.

8. Reassemble the detector and blow dry carrier gas through it before heating to operating temperatures.

9. Allow the detector to remain at operating temperatures for several hours before use.

Appendix B

Care and Handling of Microsyringes

A too often overlooked aspect of the operation of a
chromatograph is the use of the injection syringe.
This initial operation can have as much effect upon
the instrument performance as any other system. Since
microsyringes deal in very small quantities (1-10 μl),
it is essential that the syringes be kept clean.
Cleaning must be accomplished with great care since
most microsyringes are quite fragile and their accuracy can easily be destroyed. Be sure to consult
the instructions before disassembling the syringe
rather than afterwards to find out what was done
wrong. The technique used for injecting the sample
is also important. It is the primary objective of
the injection to introduce a sharp "plug" of sample
into the chromatograph inlet. Improper injection can
cause serious loss of column efficiency. However,
by following the few procedures outlined below and
with sufficient practice, most microsyringes are capable of quantitative accuracy and repeatability.

Cleaning

Always be sure to clean the syringe before and after
use to prevent contamination of samples. If high
molecular weight samples are being injected, it is
advisable to flush the syringe immediately after use
with a suitable solvent to prevent build-up of hard
to remove deposits. For general cleaning of an as-
sembled syringe, the following method will usually
be successful:

A. Flush the syringe several times with the follow-
    ing solutions in the order indicated:

    1. 5% potassium hydroxide/water
    2. Distilled water
    3. Acetone
    4. Chloroform

B. Remove remaining chloroform by heating and/or
    vacuum. A simple device for this purpose is made
    by the Hamilton Syringe Company, Whittier,
    California 90608.

Liquid Injection Technique

A. Be certain syringe is clean

B. Draw up sample into syringe and flush syringe with sample a few times.

C. Overfill syringe with sample

D. Invert syringe (needle up) to allow air to rise.

E. Discharge excess sample to desired sample volume.

F. Examine the syringe for bubbles. If they are present, repeat filling procedures. It is sometimes helpful to discharge sample rapidly into sample a few times in order to remove bubbles.

G. Lightly wipe needle with a lint-free tissue, being careful not to draw sample out of needle.

H. In a quick, smooth motion, insert syringe into septum, inject sample, and withdraw syringe.

I. Flush with solvent if necessary.

Appendix C

Troubleshooting Clues from the Rotameter

The Rotameter attached to the carrier gas line can provide clues to problems occuring in the flow system.

A. Normal Operation

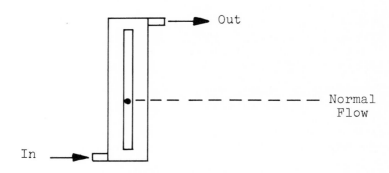

B. Flow System Faults

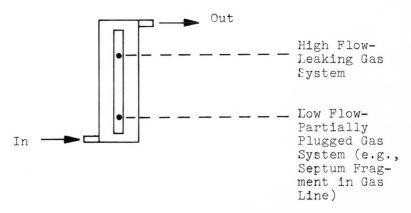

If the Rotameter ball falls below the "Normal" position during a temperature programmed analysis, a faulty flow controller is indicated. The situation may be caused by a low inlet pressure to the flow controller.

APPENDIX D

Trouble-shooting of

Silicon Controlled Rectifer Heating Systems

The oven and temperature controllers of many modern chromatographs utilize trigger pulse circuits and SCR's to control the power input to the oven heating coils. A typical block diagram of such a circuit is shown in Figure D-1.

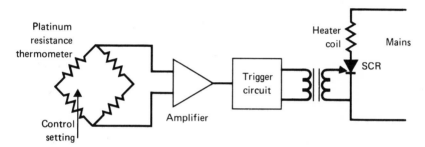

Figure D-1   A Typical Block Diagram of An
              SCR Trigger Circuit

The trigger pulse is derived from a resistance bridge, amplifier and pulse shaping circuit. A platinum resistance thermometer inside the oven and the temperature control settings form two arms of the bridge circuit. This signal couplet with a phase indicating signal, often ac from the secondary coil of a transformer, form the input to the amplifier. The amplifier output is coupled to the pulse shaping circuit which produces the SCR trigger pulse. This phase varies in phase with respect to the voltage across the SCR when the oven is in the region of the control temperature and, thus, the power input to the oven is controlled by the out of balance signal of the measuring circuit.

271

The common faults on such a circuit are when the oven either fails to heat or overheats.

An oscilloscope is invaluable for tracing faults. If the oven fails to heat or fails to heat properly the output from the SCR can be checked. If the output phase patterns Figures D-2a, 2b and 2c are not present ,then the trigger pulse to the gate of the SCR should be checked. If no trigger pulse is present then the measuring circuit and oven controller should be checked for short circuits, open circuits and loose connections. Further attention to the amplifier and trigger circuit is best left to a qualified service man.

If the trigger circuit is present but the oven still fails to heat, the SCR should be checked and replaced if faulty.

The oven overheating situation is usually caused by one of three situations. Again the oscilloscope is of great value in tracing the fault.

The first situation is caused by uncontrolled opera-tion of the SCR. This is usually a short across the SCR which must be replaced.

If the SCR is good then the problem is caused by un-controlled trigger pulses caused by a fault in the trigger generating circuit or, more likely, a shorted platinum resistance thermometer.

The third situation occurs with a phase reversal where the circuit provides more heat instead of less heat. This can lead to an uncontrolled runaway. The usual initial cause is the replacement of a plug or another connecting cable.

Occasionally an intermittent heating condition can arise. This is sometimes caused by incorrect connection of the controller to the oven power supply giving an out-of-phase condition. However, th the most frequent cause of this condition is a loose connection.

Oscilloscope set differential across heater leads:
a)   Oven temperature below set point

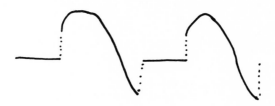

b)   Oven temperature close to set point (Normal)

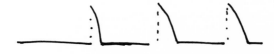

c)   Oven temperature above set point

Figure D-2   Typical Oscilloscope SCR Patterns

Appendix E

A Rapid Fault Isolation Method

Rapid isolation of faults within the GC system can be
a valuable aid in troubleshooting. The method of-
fered here utilizes two simple checking circuits and
a dummy column. Essentially, the checking circuits
and dummy column are simple, temporary replacements
for actual GC systems which might be at fault. The
first checking circuit can act as a replacement electrometer, dial in theoretical peaks, act as a replacement for hot wire and thermistor TC detectors, or
check out the operation of digital integrators. The
second checking circuit acts as a callibration device
to check for the proper operation of the strip chart
recorder. Faults can be isolated to the column or
inlet system by installing the dummy column.

A single electrometer-hotwire-thermistor detector
(EHTD) checker can be built from the wiring diagram
in Figure E-1. All resistors should be precision
and matched and the circuit mounts should be isolated
from the case. A series of resistors should also be
attached to a terminal strip to be used as replacements for TCD elements. The resistors should match
the resistances of the filaments or thermistors and
should be mounted in groups of two (for two element
detectors) or four (for four element detectors) to
allow for replacement of all detector elements
simultaneously.

The second checking circuit, the recorder checker
can be built according to the wiring diagram in
Figure E-2. By connecting this checkingcircuit directly to the recorder, it can be easily checked for
calibration. The series of switches will cause the
recorder pen to travel 90%, 10% and 100% of full
scale when they are depressed. This recorder checker
is useful in calibrating 1,5, and 10 mV recorders
to within $\pm$ 2% accuracy. Callibration adjustments
should be made according to the instrument manual.

The dummy column consists of a 10 foot length or
1/4" o.d. tubing filled with a molecular sieve
support. It is simply substituated for the actual
separation column to determine if the fault dis -
appears.

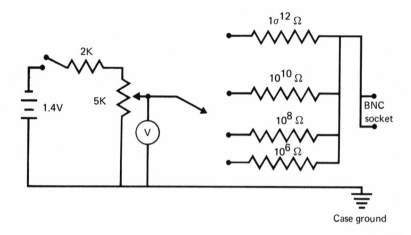

Figure E-1 Electrometer/Hot Wire/Thermister Detector
Checker Wiring Diagram

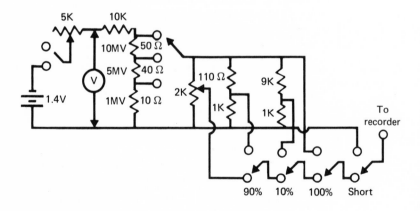

Figure E-2 Recorder Checker and Calibration Wiring
Diagram

The two checking circuits and the dummy column can
be used to isolate faults when used according the
diagram in Figure E-3. This diagram provides a logical method of isolating the fault. It should be
noted that this fault isolation method is not intended as a substitute for Chapter 14, "Gas Chromatographic Comprehensive Troubleshooting". They should rather be complimentary.

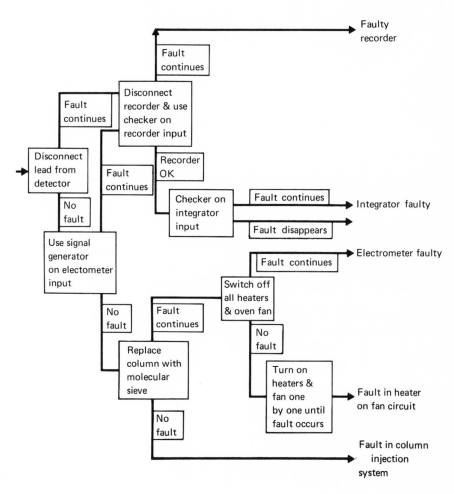

Figure E-3 Rapid Fault Isolation Diagram

Appendix F

Gas Chromatographic Column Preparation

Preparing one's own columns, rather than buying commercially prepared columns, can provide a significant
cost savings without loss of column capability.
Column preparation does not require a great deal of
equipment and the materials are inexpensive when
compared with the cost of a commercially prepared
column. Preparing GC columns does, however, require
time and care. If a significant number of columns
are needed, the investment in time can be very worthwhile. The methods presented here for preparing both
packed columns and open tubular columns have been
used successfully to prepare many columns. The
methods are not entirely fool-proof though and a certain degree of failure(relatively inefficient columns)
can be expected.

Once the column dimensions have been chosen and the
correct liquid phase determined, preparation of the
column can begin. All coating procedures are intended
to produce uniform packing without voids which will
cause losses of efficiency.

Cleaning Column Tubing

As received, tubing, particulary metallic tubing(copper and stainless steel), often contains residues
from its manufacturer. A cleaning procedure suggested by Gouw, Whittemore, and Jentoft(1) has been
quite effective for removing these residues. Cleaning
is accomplished by successive flushings with the
following reagents:

1. Methylene Chloride (CH_2CL_2)
2. Acetone (C_3H_6O)
3. Water
4. Concentrated Nitric Acid (HNO_3) or concentrated hydrochloric acid (HC_1)
5. Water
6. Ammonium Hydroxide (NH_4OH)
7. N-methyl Pyrrolidone
8. Acetone
9. Liquid phase solvent

In each case, flushing should be continued until the

reagent emerges clear. Hydrochloric acid (1) for
nitric acid since HNO_3 will react with the copper.
These reagents should be pushed through the column
with an inert carrier gas such as nitrogen or helium.
Following cleaning, the tubing should be dried with
the carrier gas stream for about twenty minutes. The
tubing should then be capped under a positive carrier gas pressure to awaiting packing.

Column Packing Preparation

The coating of the liquid phase on the inert support
material must be carried out carefully to prevent
crushing of the support material and to obtain complete, uniform coatings.

High boiling solvents are preferred when dissolving
the liquid phase. A solvent such as toluene would
be preferred to such materials as methylene chloride
or acetone (2). Using high boiling solvents will
reduce the rate of evaporation and provide more
uniform coatings.

In the evaporation technique of preparing column
packings, the liquid phase/solvent solution is
thoroughly mixed with the inert support. Stirring
should be accomplished gently with a blunt instrument
to avoid fracturing the support material. This
slurry is then poured into a shallow evaporating pan
and the solvent allowed to evaporate. The evaporation should take place slowly with constant stirring.
If the slurry is not stirred during the evaporation
process, liquid phase will migrate to the top resulting in a high concentration of liquid phase
there.

The evaporation technique is useful for 10-30%
liquid phase columns. At percentages less than a-
round 10%, the coatings must be more uniform than
this method can provide. For these low liquid loaded
columns, a filtration technique should be used.
After the slurry has been formed, it is poured into
a glass frit funnel , and the coating solution allowed to drain off. The % liquid phase will be approximately equal to the % of liquid phase in the solution. More accurate determinations can be made by
weighing the support before and after coating(3).
Once again, it is important to stir the support

constantly during drying to prevent migration of the liquid phase to the top.

Packed Column Preparation

After the column tubing is cleaned and the support m material coated, the column is ready to be prepared. Columns longer than ten feet should be prepared in increments and joined later with unions. Longer columns are difficult to handle. If the column is six feet or less, it should be coiled prior to packing to prevent crushing of the support during coiling.

Liquid phase and support are packed into the column with the aid of a vacuum pump and sometimes a vibrator. First, a plug of glass wool is inserted into the outlet end of the column. A vacuum pump is then attached to this end of the column to provide a light pull on the column packing. The glass wool plug keeps the packing in the column. Packing is poured into the column through a funnel at the column inlet. For columns which are packed uncoiled, a hand held vibrator can be used to improve the packing. The vibrator should be started at the outlet end. This can be repeated several times. After the column is completely packed, another glass wool or quartz wool plug should be inserted into the column inlet. The column is now ready for conditioning.

Coating Open Tubular Columns

Preparation of open tubular columns is somewhat more difficult than preparing packed columns even though fewer steps and less equipment is involved. The columns are prepared by pushing a coating solution through the narrow bore tubing and the difficulty arises in obtaining a thin uniform coating of liquid phase on the column walls. The procedure described here is after The Dynamic Method by Ettre (4).

A thoroughly cleaned reservoir capable of containing more than the volume of the column is needed. One end of the reservoir is attached to an inert gas supply (Nitrogen or Helium) and the other end is attached to a 6 foot x 0.01 inch i.d. pre-column restrictor. The gas supply must be equipped with accurate flow controllers and a bleed valve to help maintain constant flow through the column. The

column tubing is attached to the pre-column restrictor
at one end and another 6 foot x 0.01 in post-column restrictor at the other end. At the outlet of the post-
column restrictor, a soap bubble flow meter or a
water reservoir can be used to monitor the flow of the
coating solution.

To coat the column, a measured quantity of coating
solution in excess of the column volume is filtered
and poured into the reservoir. The pressure from the
carrier gas supply is gradually increased until the
proper flow rate is obtained. Coating solution should
be forced through the column slowly at about 2-5 cm^3/sec
linear velocity. The appropriate volumeter flow rate
should be determined for the column diameter used.
As the coating solution leaves the post-column restrictor the flow rate will begin to increase, slowly
at first then very rapidly. By reducing the inlet
pressure and using the bleed valve to release back
pressure, a fairly constant flow rate can be maintained. The coating solution should be collected at
the outlet to determine the average coating thickness
of the liquid phase (based on the total area of the
internal column walls).

After the coating solution has completely emerged
from the column, a small flow of carrier gas should
be passed through the column for one hour. The
column is now ready for conditioning.

Column Conditioning

All columns must be conditioned before use to remove
traces of coating solvents and other impurities which
might interfere with an analysis. The column should
be placed in a column oven with the inlet connected
to the carrier gas supply and the outlet left to
discharge into the atmosphere. With gas flowing
through the column, the temperature should be increases slowly. The column temperature should be
maintained somewhere above the temperature at which
it will be operated and allowed to remain at that
temperature overnight. Never condition the column
for this length of time at temperatures within 20°C
of the maximum operating temperature of the liquid-
phase. After conditioning, the column is ready for
use.

References

1. T.H. Gouw, I.M. Whittemore, and R.E. Jentoft, Anal.Chem, 42, 12 (1970).

2. Short Course on Column Selection, On Tape, Supelco, Inc., Bellefonte, Penna., 1971

3. S.F. Spencer in Instrumentation in Gas Chromatography, J. Krugers, Ed., Centrey Publishing Company (eindhoven), 1968.

4. L.S. Ettre, Open Tubular Columns in Gas Chromatography, Plenum Press (New York), 1965.

Appendix G

Glossary of Terms and Formulas

Absolute Retention Time: (t_R) See, Retention Time

Absolute Retention Volume: (V_R) See, Retention Volume

Adjusted Retention Time: (t_R') Absolute retention
time (q.v., retention time) minus retention time
of a substance which does not interact with the
stationary phase (i.e., air, for many columns).
Units: usually in millimeters of chart, through
it can be given in units of time.

Adjusted Retention Volume: (V_R') Absolute retention
volume (q.v., retention volume) minus gas holdup.
Units: cm^3

Adsorbent: Stationary phases which interact with
sample components only at the surface of the phase.

Air Peak: The peak of any substance which does not
interact with the stationary phase. (for many
columns this will be air).

Attenuator: The part of the electrical system which
determines the amount of amplification of detector
signals being sent to the recorder.

Baseline: The straight line drawn by the recorder
pen when no sample is entering the detector.

Capillary Column: A small inner diameter column
(0.25-1.0mm) which contains a stationary phase
attached to column walls either directly or on
a solid support.

Carrier Gas: An inert gas which pushes the sample
through the column. Common carrier gases are
helium, nitrogen, argon, and argon/methane.

Carrier Gas Velocity: (μ) Mass flow rate of the
carrier gas. Units: ml/min.

Chromatogram: The series of peaks produced by the
strip chart recorder representing the sample.

Column: The tubing (metal, glass, or teflon) containing the stationary phase.

Detector: The device which converts the presence of the sample component at the end of the column into an electrical signal. The detector monitors changes in some physical property of the gas.

Elution: The emergence of a component from the column.

Efficiency: (N) The ability of a column to maintain sharp (narrow) peaks. It is the primary criteria for judging column quality.

$$N = 16 \left(\frac{t_R'}{w}\right)$$

t_R', adjusted retention time (cm)
w, peak width at base (cm)

Units: Number of theoretical plates

Flow Programming: Predetermined increase in carrier gas flow rate to expedite elution of high boiling components without raising column temperature.

Gas Holdup: (Vm) Volume of carrier gas which passes through the column in order to elute a substance which does not interact with the stationary phase.

Gas- Solid Chromatography: Gas chromatography in which the stationary phase is a solid adsorbent.

Ghost Peaks: Unexpected chromatogram peaks usually caused by contamination of the GC system or the sample.

Height Equivalent to a Theoretical Plate: Column efficiency as related to column length.

$$HETP = \frac{L}{N}$$

L, column length (cm)
N, efficiency

Note: The reciprocal of HETP (1/HETP=N/L) is frequently used to describe efficiency in plates per unit length.

Integrator: An instrument used to determine the area
    under a chromatographic peak. The measurement
    is either made mechanically, or it is made elect-
    ronically providing a continuous digital read-out.

Ionization Detectors: Detectors which measure cur-
    rent produced by the migration of ionized parti-
    cles in the carrier gas stream. The particles are
    ionized usually by either heat or by a radio-
    active source.

Katharometer: See, thermal conductivity detector.

Leading Peaks: Peaks characterized by a gradual rise
    of the recorder pen followed by a rapid return
    to base-line.

Mean Carrier Gas Velocity: ($\bar{\mu}$) Carrier gas velocity
    over the length of the column.

Mobile Phase: The carrier gas which pushes sample
    components through the column.

Open Tubular Columns: See, capillary columns.

Pyrolysis: The thermal degradation of a sample be-
    fore introduction into the column.

Relative Retention Time: (r) A measure of the capa-
    bility of a stationary phase to separate two com-
    ponents. It is the ratio of the adjusted retent-
    ion times for the two components.

$$r_{2-1} = \frac{t_{R'}}{t_{R'}}$$

Note: r is also known as the Separation Factor(α)

Resolution: (R) The degree to which two peaks are
    separated. It is a function of peak with and
    distance between the peaks.

$$R = \frac{V_{R_2} - V_{R_1}}{w_1 + w_2}$$

When R=1.5, separation is complete.

Retention Time: (t_R) The time required for a substance to pass through the column. It is measured
from the time of injection of the sample to the
time at which the top of the peak appears. Units:
cm or units of time.

Retention Volume: (V_R) The product of the retention
time (min) and flow rate (ml/min.). Units: ml

Rotometer: Device for measuring gas flow rates consisting of one or more small balls contained in a
length of glass tubing. The flow rate is a function of the height of the ball in the tubing.

Selectivity: The tendency of a detector to respond
to certain classes of compounds to a greater degree than other classes of compounds.

Separation Factor: (a) See, Relative retention time.

Soap Bubble Meter: A device for accurately measuring low flow rates. It consists of a volumetric
glass tube with a soap reservoir through which
the gas is flowing. The flow rate is determined
by the time required for a soap bubble to move
between calibration marks.

Solid Support: A solid material upon which the
liquid phase is coated.

Solvent: As applies to GC, it generally refers to
the liquid phase. It may also indicate the substance used to dilute or transport the sample.

Specificity: See, selectivity.

Splitter: A fixed or variable device for reducing
the quantity of sample entering the column from
the inlet.

Stationary Phase: The liquid or solid material in
the column which causes separation to occur.

Tailing: A peak characterized by a rapid rise to
its peak, followed by a gradual return to the
baseline.

Temperature Programming: Increasing column

temperature at a predetermined rate to enhance separation and speed elution of high boiling components.

Theroretical Plates: (N) Unit of column efficiency (q.v.) Height Equivalent to a theoretical plate.

Thermal Conductivity Detector: A GC detector which measures changes in the conductivity of the carrier gas caused by the presence of sample components.

Vapor-Phase Chromatography: Synonomous with Gas Chromatography.